PUBLISHER COMMENTARY

We print NASA's handbooks and standards for the convenience of those that use them on a daily basis. We print all of these a full 8 ½ by 11 with large text so they are easy to read. Yes, color books are expensive to print so unless the information relies on the use of color for proper interpretation or understanding, we print most books in black and white to keep the cost down. All these documents are available for download for free from NASA, however printing them all over a network printer would take days.

Why buy a book you can download free? We print this so you don't have to.

All these books are available for free download from the government web site. Some are available only in electronic media. Some online docs are missing pages or barely legible.

We at 4th Watch Publishing are former government employees, so we know how government employees actually use the standards. When a new standard is released, an engineer prints it out, punches holes and puts it in a 3-ring binder. While this is not a big deal for a 5 or 10-page document, many NIST documents are over 100 pages and printing a large document is a time-consuming effort. So, an engineer that's paid $75 an hour is spending hours simply printing out the tools needed to do the job. That's time that could be better spent doing engineering. We publish these documents so engineers can focus on what they were hired to do – engineering. It's much more cost-effective to just order the latest version from Amazon.com

If there is a standard you would like published, let us know. Our web site is www.usgovpub.com

www.usgovpub.com

List of Other NASA Publications Available on Amazon.com:

NASA-STD-5001B	Structural Design and Test Factors of Safety for Spaceflight Hardware
NASA-STD-5006A	General Welding Requirements for Aerospace Materials
NASA-STD-5008B	Protective Coating of Carbon Steel, Stainless Steel, and Aluminum on Launch Structures, Facilities, and Ground Support Equipment
NASA-STD-5009A	Nondestructive Evaluation Requirements for Fracture-Critical Metallic Components
NASA-STD-5012B	Strength and Life Assessment Requirements for Liquid-Fueled Space Propulsion System Engines
NASA-STD-5019A	Fracture Control Requirements for Spaceflight Hardware
NASA-STD-5005D	Standard for The Design and Fabrication of Ground Support Equipment
NASA-HDBK-8739.21	Workmanship Manual for Electrostatic Discharge Control
NASA-HDBK 8739.23A	NASA Complex Electronics Handbook for Assurance Professionals (Color)
NASA-HDBK-8719.14	Handbook for Limiting Orbital Debris (Color)
NASA-HDBK-8709.22	Safety and Mission Assurance Acronyms, Abbreviations, and Definitions
NASA-HDBK-7009	NASA Handbook for Models and Simulations: An Implementation Guide For NASA-STD-7009 (Color)
NASA-HDBK-8739.19-2	Measuring and Test Equipment Specifications NASA Measurement Quality Assurance Handbook – Annex 2
NASA-HDBK-8739.19-3	Measurement Uncertainty Analysis Principles and Methods NASA Measurement Quality Assurance Handbook – Annex 3
NASA-HDBK-8739.19-4	Estimation and Evaluation of Measurement Decision Risk NASA Measurement Quality Assurance Handbook – Annex 4
NASA RCM	Reliability-Centered Maintenance Guide for Facilities and Collateral Equipment

www.usgovpub.com

METRIC/SI (ENGLISH)

NASA TECHNICAL STANDARD

National Aeronautics and Space Administration

NASA-STD-5008B w/CHANGE 1: REVALIDATED w/ ADMINISTRATIVE/ EDITORIAL CHANGES 2016-05-31

Approved: 2011-03-18
Superseding NASA-STD-5008A

PROTECTIVE COATING OF CARBON STEEL, STAINLESS STEEL, AND ALUMINUM ON LAUNCH STRUCTURES, FACILITIES, AND GROUND SUPPORT EQUIPMENT

DOCUMENT HISTORY LOG

Status	Document Revision	Change Number	Approval Date	Description
Baseline			07-10-2001	Baseline Release.
Revision	A		01-21-2004	See changes below.

Foreword: Changed signature of W. Brian Keegan to Theron M. Bradley. Updated Table of Contents and KSC FORM 21-61ONS, Standardization Document Improvement Proposal.

General editorial/punctuation changes/corrections throughout document: put brackets inside parentheses and deleted "etc." following "e.g." Used non, free and proof as combining forms. Put a hyphen between Sherwin Williams. Changed Ameron International, P.C.G. to P.C.F.G. and web site to www.ameron.com. Changed Materials Science Division (LO-G) to Spaceport Technology Development Office (YA-C2).

Added paragraph 1.6, Environmental stewardship.

Paragraph 2.2: Changed MIL-C-24667A to MIL-PRF-24667A. Deleted MIL-T-81772.

Paragraph 2.3: Changed title of ASTM A653 to Standard Specification for Steel Sheet, Zinc-Coated (Galvanized) or Zinc-Iron Alloy-Coated (**Galvannealed**) by the Hot-Dip Process and changed the ASTM address to 100 Barr Harbor Drive, West Conshohocken, PA 19482-2959. Changed the title of RP0188-88 to Discontinuity (Holiday) Testing of New Protective Coatings on Conductive Substrates. Changed the title of SSPC SP 5-94, Joint Surface Preparation Standard, to SSPC-SP 5-94/NACE No. 1, White Metal Blast Cleaning. Changed SSPC SP 10-94, Joint Surface Preparation Standard Near-White Blast Cleaning, to SSPC-SP 10-00/NACE No. 2, Near-White Blast Cleaning.

Paragraph 3: Deleted AISC, American Institute of Steel Construction; cfm, cubic feet per minute; GFE, Government-furnished equipment; and KHB, Kennedy handbook. Changed LO-G to YA-C2-T. Added ASME, American Society of Mechanical Engineers; CFR, Code of Federal Regulations; HAP, hazardous air pollutant; PDCA, Painting and Decorating Contractors of America; and YA-F, Labs and Testbed Division. Changed mega Pascal to megapascal.

Paragraph 4.1.2.4.e.2: Changed 235 to 239. Put parentheses around number listing.

Paragraph 4.1.4: Changed Materials and Chemical Analysis Laboratory to Spaceport Technology Development Office.

Paragraph 4.3.1: Added "and NASA policies" to the end of the last sentence.

Status	Document Revision	Change Number	Approval Date	Description
Revision (Continued)	A		01-21-2004	See changes below.
	4.4.1: Rewrote the first sentence. Added a sentence after the first sentence. 4.4.3.3 Added/environmental to safety/fire in the last sentence. 4.5.1: Added "An inorganic zinc coating used in a friction-type joint must be approved by the American Institute of Steel Construction (AISC)," after the fifth sentence. Paragraph 5.7.5, end of the third sentence: Added "while considering the accuracy of the measurement instrument." Appendix A, Section I: Changed zip code for Ameron International, P.C.F.G to 92821. Appendix A, Section II: Added Dimetcote D-9HS SB; SB to Cathacoat 304V; Cathacoat 304H SB; and ZincClad IIHS SB to Coating Designations and Type. Added InterZinc 22HS, SB, International Paint, 6001 Antoine Drive, Houston, TX 77091, (713) 682-1711, www.international-pc.com. Appendix B, Section II: Added (SB) to Primer (Type) D-21-9. Added D-9HS (SB) to Primer (Type) and PSX700(SB) to Topcoat (Type). Appendix B, Section III: Added D-9HS (SB) and ZincClad IIHS (SB) to Primer (Type). Added InterZinc 22HS (SB), 181 (SB) (IOT), International Paint, 6001 Antoine Drive, Houston, TX 77091, (713) 682-1711, www.international-pc.com. Changed Topcoat (Type) 7551 P1 (SB) (IOT) to 5555.			
Revision	B		03/18/2011	General Revision: Changes are not listed; reader must review complete document.
		1	2016-05-31	Revalidated w/Administrative/Editorial Changes—This NASA Technical Standard was reviewed and no technical changes resulted, but some requirements were clarified. Administrative changes to number requirements, add a Requirements Compliance Matrix as Appendix A, and conform to the current template were made, along with editorial corrections.

FOREWORD

This NASA Technical Standard is published by the National Aeronautics and Space Administration (NASA) to provide uniform engineering and technical requirements for processes, procedures, practices, and methods that have been endorsed as standard for NASA programs and projects, including requirements for selection, application, and design criteria of an item.

This NASA Technical Standard is approved for use by NASA Headquarters and NASA Centers and Facilities and may be cited in contract, program, and other Agency documents as a technical requirement. It may also apply to the Jet Propulsion Laboratory and other contractors only to the extent specified or referenced in applicable contracts.

This NASA Technical Standard was developed to ensure the inclusion of essential criteria in the coating of ground support equipment (GSE) and facilities used by or for NASA.

This NASA Technical Standard establishes practices for the protective coating of GSE and related facilities used by or for NASA programs and projects. This NASA Technical Standard is for the design of nonflight hardware used to support the operations of receiving, transportation, handling, assembly, inspection, test, checkout, service, and launch of space vehicles and payloads at NASA launch, landing, or retrieval sites. These criteria and practices may be used for items used at the manufacturing, development, and test sites upstream of the launch, landing, or retrieval sites.

The information provided herein is recommended for use in the preparation of written, individual coating specifications for specific projects for the prevention of corrosion through the use of protective coatings on facilities, space vehicle launch structures, and GSE in all environments.

Requests for information should be submitted via "Feedback" at https://standards.nasa.gov. Requests for changes to this NASA Technical Standard should be submitted via MSFC Form 4657, Change Request for a NASA Engineering Standard.

Original Signed By:	*03/18/2011*
Michael G. Ryschkewitsch NASA Chief Engineer	Approval Date

TABLE OF CONTENTS

TABLE OF CONTENTS (Continued)

TABLE OF CONTENTS (Continued)

LIST OF APPENDICES

APPENDIX **PAGE**

LIST OF TABLES

TABLE **PAGE**

PROTECTIVE COATING OF CARBON STEEL, STAINLESS STEEL, AND ALUMINUM ON LAUNCH STRUCTURES, FACILITIES, AND GROUND SUPPORT EQUIPMENT

1. SCOPE

1.1 Purpose

1.1.1 The purpose of this NASA Technical Standard is to establish requirements for the application of protective coatings to mitigate corrosion of exposed carbon steel, stainless steel, and aluminum.

1.1.2 This NASA Technical Standard provides a design standard for experienced corrosion control engineers for the development of specifications, including requirements for materials, equipment, safety, procedures, and quality assurance inspections.

1.1.3 Refer to section 1.2 for the intended use of this NASA Technical Standard and surfaces to be coated according to this NASA Technical Standard. Refer to Appendices A, B, and C for listings of approved coating materials.

1.2 Applicability

1.2.1 This NASA Technical Standard is applicable to facilities, launch structures, ground support equipment (GSE), test facilities, and structures that are intended for use at all NASA locations world-wide.

1.2.2 This NASA Technical Standard is approved for use by NASA Headquarters and NASA Centers and Facilities and may be cited in contract, program, and other Agency documents as a technical requirement. It may also apply to the Jet Propulsion Laboratory and other contractors only to the extent specified or referenced in applicable contracts.

1.2.3 Verifiable requirements are numbered and indicated by the word "shall"; this NASA Technical Standard contains 333 requirements. Explanatory or guidance text is indicated in italics beginning in section 4. To facilitate requirements selection and verification by NASA programs and projects, a Requirements Compliance Matrix is provided in Appendix J.

1.2.4 [PCR 1] This NASA Technical Standard shall be used in the preparation of written, individual coating specifications for specific projects for the prevention of corrosion through the use of protective coatings on space vehicle launch structures, facilities, GSE, and test facilities and structures in the specific environments identified in section 1.4.

1.2.5 [PCR 2] Due to the changing environmental considerations and different site conditions, new advances in corrosion technology, and a wide array of possible applications, this NASA Technical Standard shall not be used as a stand-alone specification that meets every contingency.

1.2.6 [PCR 3] The appendices are considered to be an integral part of this NASA Technical Standard. Appendices A, B, C, and D shall be used for the preparation of all coating specifications.

1.3 Tailoring

[PCR 4] Tailoring of this NASA Technical Standard for application to a specific program or project shall be formally documented as part of program or project requirements and approved by the Technical Authority in accordance with NPR 7120.5, NASA Space Flight Program and Project Management Requirements.

1.4 Zones of Exposure

[PCR 5] The zones of exposure established to define coating system requirements for surfaces located in specific environments shall be determined by the Design Engineer responsible for preparing the coating specification from the following zones:

a. (1) Zone 1a. Surfaces that are directly impinged on by solid rocket booster (SRB) engine exhaust.

(2) Zone 1b. Surfaces that are indirectly impinged on by SRB exhaust.

(3) Zone 1c. Walking surfaces in Zones 1a and 1b.

b. Zone 2. Surfaces that are exposed to elevated temperatures (above 65 °C (above 150 °F)) and/or acid deposition from SRB exhaust with no exhaust impingement.

c. (1) Zone 3a. Surfaces, other than those located in Zones 1 or 2, that are exposed to acid deposition from SRB exhaust products.

(2) Zone 3b. Surfaces that are exposed to other types of chemical contamination (e.g., cooling towers, diesel exhaust stacks, acidic industrial environments, and water treatment facilities).

d. (1) Zone 4a. Surfaces not located in the launch environment but located in a neutral pH corrosive marine industrial environment or other chloride-containing environments.

(2) Zone 4b. Surfaces located in neutral pH exterior environments in any geographical area.

(3) Zone 4c. Surfaces located in indoor environments that are not air-conditioned.

e. (1) Zone 5a. Surfaces located in a continuous indoor air-conditioned environment, such as an office or clean room, where both temperature and humidity are controlled more than 90 percent of the time.

(2) Zone 5b. Surfaces located in a low humidity, high ultraviolet environment, such as a high altitude, arid location.

f. (1) Zone 6a. Surfaces located underground or subject to intermittent or continuous immersion in aqueous environments.

(2) Zone 6b. Surfaces subject to exposure in a chemical/fuel storage environment. Based on the complexity of the liquid stored, this has to be engineered separately in compliance with all federal, state, and local environmental statutory requirements.

g. Zone 7. Surfaces under thermal insulation, such as chilled water, steam, and heated gas lines.

1.5 Method of Specifying Coating Requirements

a. [PCR 6] Specifications referencing this NASA Technical Standard shall include the following:

(1) The type of surface to be coated.
(2) The zone of exposure.
(3) Surface preparation.
(4) Defined paint system.
(5) Coating thicknesses.
(6) The finish color required (when applicable).

These requirements should be assembled in a coating schedule for easy reference.

b. [PCR 7] The coating specification shall contain the following key elements: scope, applicable documents, submittals, environmental protection, waste management, safety/personnel protection, materials, tools and equipment, environmental conditions, work schedule, surface preparation (including a listing of abrasive-sensitive hardware to be prepared or protected), coating schedule, coating mixing and application, quality control inspection, reporting, and final acceptance.

See Appendix D, Coating Specification Key Elements, for a recommended outline of a coating specification and Appendix E for a recommended coating schedule.

1.6 Environmental Stewardship and Health and Safety

a. [PCR 8] Environmental, health, and safety impacts of processes and materials shall be taken into account when employing protective coating methods and techniques.

b. [PCR 9] Alternative, environmentally friendly materials that do not contain hexavalent chromium, lead, cadmium, or hazardous air pollutants (HAPs), such as methyl ethyl ketone, toluene, and xylene, shall be considered when determining the correct coating method/technique for each protective coating application.

c. [PCR 10] Alternative, less hazardous materials shall be considered when determining the correct coating method/technique for each protective coating application to minimize risk in construction, use, and demolition.

d. [PCR 11] Performance criteria defined in this NASA Technical Standard shall take precedent. Coatings containing hazardous materials may be harmful to human health and the environment.

2. APPLICABLE DOCUMENTS

2.1 General

The documents listed in this section contain provisions that constitute requirements of this NASA Technical Standard as cited in the text.

2.1.1 [PCR 12] The latest issuances of cited documents shall apply unless specific versions are designated.

2.1.2 [PCR 13] Non-use of specifically designated versions shall be approved by the responsible Technical Authority.

The applicable documents are accessible at http://standards.nasa.gov, may be obtained directly from the Standards Developing Body or other document distributors, or information for obtaining the document is provided.

2.2 Government Documents

Department of Defense (DoD)

MIL-A-22262	Abrasive Blasting Media Ship Hull Blast Cleaning
MIL-PRF-24667	Coating System, Non-skid, for Roll, Spray, or Self-Adhering Application

QPL 22262	Qualified Products List (Military) of Products Qualified Under Detail Specification MIL-A-22262 Abrasive Blasting Media Ship Hull Blast Cleaning

Federal (FED)

29 CFR 1910.134	Respiratory Protection
FED-STD-595	Colors Used in Government Procurement

National Aeronautics and Space Administration (NASA)

NPR 7120.5	NASA Space Flight Program and Project Management Requirements
KSC-SPEC-F-0006	Heat and Blast Protection Coating Materials and Application Methods, Specification for
KSC-STD-SF-0004	Ground Piping Systems Color Coding and Identification, Safety Standard for
NASA-STD-6001	Flammability, Offgassing, and Compatibility Requirements and Test Procedures

(Copies of specifications, standards, drawings, and publications required by suppliers in connection with specific procurement functions should be obtained from the procuring activity or as directed by the Contracting Officer.)

2.3 Non-Government Documents

American Society for Testing and Materials (ASTM)

ASTM A123	Standard Specification for Zinc (Hot-Dip Galvanized) Coatings on Iron and Steel Products
ASTM A153	Standard Specification for Zinc Coating (Hot-Dip) on Iron and Steel Hardware
ASTM A653	Standard Specification for Steel Sheet, Zinc-Coated (Galvanized) or Zinc-Iron Alloy-Coated (Galvannealed) by the Hot-Dip Process
ASTM A780	Standard Practice for Repair of Damaged and Uncoated Areas of Hot-Dip Galvanized Coatings

ASTM C920	Standard Specification for Elastomeric Joint Sealants
ASTM D520	Standard Specification for Zinc Dust Pigment
ASTM D610	Standard Practice for Evaluating Degree of Rusting on Painted Steel Surfaces
ASTM D714	Standard Test Method for Evaluating Degree of Blistering of Paints
ASTM D1654	Standard Test Method for Evaluation of Painted or Coated Specimens Subjected to Corrosive Environments
ASTM D4285	Standard Test Method for Indicating Oil or Water in Compressed Air
ASTM D4417	Standard Test Methods for Field Measurement of Surface Profile of Blast Cleaned Steel
ASTM D4752	Standard Practice for Measuring MEK Resistance of Ethyl Silicate (Inorganic) Zinc-Rich Primers by Solvent Rub

NACE International

RP0198	Control of Corrosion Under Thermal Insulation and Fireproofing Materials – A Systems Approach
RP0288	Inspection of Linings on Steel and Concrete
SP0188	Discontinuity (Holiday) Testing of New Protective Coatings on Conductive Substrates

National Sanitation Foundation (NSF)

NSF 61	Drinking Water System Components – Health Effects

The Society for Protective Coatings (SSPC)

SSPC-AB 1	Mineral and Slag Abrasives
SSPC-AB 2	Cleanliness of Recycled Ferrous Metallic Abrasives
SSPC-AB 3	Ferrous Metallic Abrasive
SSPC-PA 2	Procedure for Determining Conformance to Dry Coating Thickness Requirements

SSPC-SP 1	Solvent Cleaning
SSPC-SP 3	Power Tool Cleaning
SSPC-SP 5/ NACE No. 1	White Metal Blast Cleaning
SSPC-SP 10-/ NACE No. 2	Near-White Metal Blast Cleaning
SSPC-SP 11	Power-Tool Cleaning to Bare Metal
SSPC-TU 4	Field Methods for Retrieval and Analysis of Soluble Salts on Substrates

2.4 Order of Precedence

2.4.1 The requirements and standard practices established in this NASA Technical Standard do not supersede or waive existing requirements and standard practices found in other Agency documentation.

2.4.2 [PCR 14] Conflicts between this NASA Technical Standard and other requirements documents shall be resolved by the responsible Technical Authority.

3. ACRONYMS AND DEFINITIONS

3.1 Acronyms, Abbreviations, and Symbols

° C	degree Celsius
° F	degree Fahrenheit
µg	Microgram
µm	Micrometer
AISC	American Institute of Steel Construction
ANSI	American National Standards Institute
APL	Approved Products List
ASME	American Society of Mechanical Engineers
ASTM	American Society for Testing and Materials
CFR	Code of Federal Regulations
CGA	Compressed Gas Association
CIP	Coating Inspector Program
cm	Centimeter
DFT	dry film thickness
DoD	Department of Defense
FED	Federal
ft	Foot

GSE	ground support equipment
GU	gloss units
HAP	hazardous air pollutant
hr	Hour
in	Inch
IOT	inorganic topcoat
kip	1,000 pounds-force
km	Kilometer
kPa	Kilopascal
KSC	John F. Kennedy Space Center
ksi	kip per square inch
LLC	Limited Liability Company
m	meter
mi	mile
mil	one thousandth of an inch
MIL	Military
mm	millimeter
MPa	megapascal
mph	miles per hour
NACE	NACE International, formerly known as National Association of Corrosion Engineers
NASA	National Aeronautics and Space Administration
NIOSH	National Institute of Occupational Safety and Health
No.	number
NSF	National Sanitation Foundation
OSHA	Occupational Safety and Health Administration
oz	ounce
PCB	polychlorinated biphenyl
PCR	Protective Coating Requirement
PDCA	Painting and Decorating Contractors of America
pH	measure of acidity or alkalinity of a solution
PPE	personal protective equipment
PPG	Pittsburgh Paint and Glass
psi	pound per square inch
QPL	Qualified Products List
RH	relative humidity
SB	solvent-based
SRB	solid rocket booster
SSPC	The Society for Protective Coatings
STD	standard
TM	Technical Manual
TO	technical order
TSC	thermal spray coating
VOC	volatile organic content
WB	water-based

WFT wet film thickness

3.2 Definitions

High-gloss Finish: A minimum of 85 gloss units (GUs) at a 60-degree angle.

Semi-gloss Finish: 60 GU to 85 GU at a 60-degree angle.

4. REQUIREMENTS

4.1 Materials

4.1.1 Abrasive-Blasting Aggregate

a. [PCR 15] Blasting aggregates shall be approved materials in accordance with MIL-A-22262, Abrasive Blasting Media Ship Hull Blast Cleaning; or SSPC-AB 1, Mineral and Slag Abrasives, Type I or II, Class A; or steel grit in accordance with SSPC-AB 3, Ferrous Metallic Abrasive, Class 1.

b. [PCR 16] Only materials approved in Qualified Products List (QPL) 22262, Qualified Products List: List of Products Qualified Under Military Specification MIL-A-22262, shall be used.

c. [PCR 17] The abrasive grade selected shall produce the required surface profile and possess physical properties that are compatible with the requirements of this NASA Technical Standard.

d. [PCR 18] The new steel grit shall be a neutral pH (6.0 to 8.0), rust-free and oil-free, dry, commercial-grade blasting grit with a hardness of 40 to 50 Rockwell C.

e. [PCR 19] Recycled steel grit shall be in accordance with SSPC-AB 2, Cleanliness of Recycled Ferrous Metallic Abrasives.

f. [PCR 20] The size shall be selected to produce the required anchor profile.

For paint removal or cleaning of aluminum, stainless steel, and fiberglass, plastic media in accordance with MIL-P-85891, Plastic Media for Removal of Organic Coatings, may be used as an alternate.

g. [PCR 21] Only aggregates that are free of crystalline silica shall be selected for use at NASA unless exemptions to this policy are coordinated with the local Occupational Health Office.

h. [PCR 22] Blasting aggregate for abrasion-sensitive hardware (such as bellows, gimbal joints, and other thin-walled components) shall be materials that do not change the surface profile.

(1) [PCR 23] Blasting operations shall not produce holes, cause distortion, remove metal, or cause thinning of the substrate.

4.1.2 Protective Coatings, Thinners, and Cleaners

The following paragraphs establish minimum requirements for each generic type of protective coating specified in this NASA Technical Standard. See section 4.4.3.1 for coating intercoat compatibility requirements.

a. [PCR 24] All coatings shall possess physical properties and handling characteristics that are compatible with the application requirements of this NASA Technical Standard.

b. [PCR 25] All coatings shall be self-curing.

c. [PCR 26] Thinners and cleaners for each coating shall be procured from the manufacturer of the coating.

d. [PCR 27] Procurement awards for coatings to be supplied according to this NASA Technical Standard shall be made only for those products in the Approved Products List (APL).

e. [PCR 28] Application characteristics shall be judged acceptable prior to beach exposure testing.

f. [PCR 29] Protective coatings shall be compatible with fluids expected in the areas to the extent required to prevent fire, explosion, or damage to facility, hardware, and GSE.

g. [PCR 30] All coating materials, when used in areas where exposure to hypergolic propellants could occur, shall be compatible with the propellants in accordance with NASA-STD-6001, Flammability, Offgassing, and Compatibility Requirements and Test Procedures.

Interested parties should be aware of this requirement and are urged to arrange for testing of their product so that they may be eligible for award of contracts or orders for coatings to be supplied in accordance with this NASA Technical Standard. To arrange for product testing and the testing criteria, manufacturers must contact the Engineering Directorate, NASA, John F. Kennedy Space (KSC) Center, FL 32899 or contact the Corrosion Technology Laboratory at http://corrosion.ksc.nasa.gov.

4.1.2.1 Inorganic Zinc Coatings

Inorganic zinc coatings that have been approved are listed in Appendix A, Approved Products List for Inorganic Zinc Coatings.

a. [PCR 31] To be listed, a coating shall meet the following minimum requirements:

(1) Self-curing, multiple-component.

(2) Dry-temperature resistance to 400 °C (750 °F) for 24 hours.

(3) Minimum shelf life of 6 months when stored in accordance with manufacturer's instructions.

(4) Minimum of 83 percent zinc by weight in the applied dry film.

(5) Contain Type III zinc dust pigment in accordance with ASTM D520, Standard Specification for Zinc Dust Pigment, and be asbestos-free, polychlorinated biphenyl (PCB)-free, lead-free, cadmium-free, and chromate-free (less than 0.002 percent by weight of mixed coating).

(6) Attain a rating of not less than 9 in accordance with ASTM D610, Standard Practice for Evaluating Degree of Rusting on Painted Steel Surfaces; and ASTM D1654, Standard Test Method for Evaluation of Painted or Coated Specimens Subjected to Corrosive Environments, when applied to composite carbon steel test panels and exposed at the KSC Beach Corrosion Test Site for the following periods:

A. 18 months for initial acceptance.
B. 5 years for final acceptance.

b. [PCR 32] Application characteristics shall be judged acceptable prior to beach testing.

4.1.2.2 Primer and/or Intermediate Coatings

These coatings are listed in Appendix B, Approved Products List for Topcoat Systems.

4.1.2.2.1 Inhibitive Polyamide Epoxy Coatings

[PCR 33] Polyamide epoxy coatings shall conform to the following minimum requirements:

a. Polyamide-cured.

b. Rust-inhibitive.

c. PCB-free, lead-free, cadmium-free, and chromate-free (less than 0.002 percent by weight of mixed coating).

d. Suitable as a primer for carbon steel, galvanized steel, and aluminum.

e. Suitable as an intermediate coat between an inorganic zinc primer and an aliphatic polyurethane finish coat.

f. Meet the compatibility requirements of section 4.4.3.1.

g. Contain a minimum of 40 percent solids by volume.

4.1.2.2.2 Noninhibitive Polyamide Epoxy Coatings

[PCR 34] Polyamide epoxy coatings shall conform to the following minimum requirements:

a. Polyamide-cured.

b. PCB-free, lead-free, cadmium-free, and chromate-free (less than 0.002 percent by weight of mixed coating).

c. Suitable as an intermediate coat between inorganic zinc primer and an aliphatic polyurethane finish coat.

d. Meet the compatibility requirements of section 4.4.3.1.

e. Contain a minimum of 40 percent solids by volume.

f. Not to be used as a primer on carbon steel.

4.1.2.2.3 Water-Reducible Intermediate Coatings

[PCR 35] Water-reducible intermediate coatings shall conform to the following minimum requirements:

a. Self-curing, one or two packages, water reducible.

b. PCB-free, lead-free, cadmium-free, and chromate-free (less than 0.002 percent by weight of mixed coating).

c. Suitable as an intermediate coat between inorganic zinc primers and water-reducible topcoats.

d. Meet the compatibility requirements of section 4.4.3.1.

e. Contain a minimum of 30 percent solids by volume.

f. Not to be used as a primer on steel.

4.1.2.3 Finish Coatings

4.1.2.3.1 Aliphatic Polyurethane Coatings

[PCR 36] Aliphatic polyurethane coatings shall conform to the following minimum requirements:

a. Catalyst isocyanate cured.

b. High-gloss finish (minimum 85 gloss units (GUs) at a 60-degree angle).

c. Retain gloss and color upon prolonged exterior exposure.

d. Suitable as an exterior finish coat over an inorganic zinc primer with a polyamide epoxy intermediate coat.

e. Meet the compatibility requirements of section 4.4.3.1.

f. Contain a minimum of 44 percent solids by volume.

g. PCB-free, lead-free, cadmium-free, and chromate-free (less than 0.002 percent by weight of mixed coating).

h. Attain a numerical rating of not less than 8 in accordance with ASTM D610 and ASTM D1654 and a numerical rating of not less than 9F in accordance with ASTM D714, Standard Test Method for Evaluating Degree of Blistering of Paints, when applied as a system to composite carbon steel test panels and exposed at the KSC Beach Corrosion Test Site for the following periods:

(1) 18 months for initial acceptance.
(2) 5 years for final acceptance.

4.1.2.3.2 Water-Reducible Topcoats

[PCR 37] Water-reducible topcoats shall conform to the following minimum requirements:

a. Self-curing, one or two packages, water-reducible.

b. PCB-free, lead-free, cadmium-free, and chromate-free (less than 0.002 percent by weight of mixed coating).

c. Retain gloss and color upon prolonged exterior exposure.

d. Semi-gloss or high-gloss finish. (Semi-gloss is defined as 60 GU to 85 GU at a 60-degree angle; high gloss is defined as a minimum 85 GU at a 60-degree angle.)

e. Meet the compatibility requirements of section 4.4.3.1.

f. Attain a numerical rating of not less than 8 in accordance with ASTM D610 and ASTM D1654 and a numerical rating of not less than 9F in accordance with ASTM D714, when applied as a system to composite carbon steel test panels and exposed at the KSC Beach Corrosion Test Site for the following periods:

(1) 18 months for initial acceptance.
(2) 5 years for final acceptance.

4.1.2.3.3 Inorganic Topcoats (IOTs)

[PCR 38] IOTs shall conform to the following minimum requirements:

a. Dry-temperature resistance to 400 °C (750 °F) for 24 hours.

b. Suitable as a topcoat for inorganic zinc and galvanized steel in high-temperature environments.

c. Listed as an approved coating system (see Appendix B).

d. PCB-free, lead-free, cadmium-free, and chromate-free (less than 0.002 percent by weight of mixed coating).

e. Attain a rating of not less than 9 in accordance with ASTM D610 and ASTM D1654 when applied to composite carbon steel test panels and exposed at the KSC Beach Corrosion Test Site for the following periods:

(1) 18 months for initial acceptance.
(2) 5 years for final acceptance.

4.1.2.3.4 Polysiloxane Topcoats

[PCR 39] Polysiloxane topcoats shall conform to the following minimum requirements:

a. Suitable as a finish coat for exterior exposure.

b. Contain a minimum of 44 percent solids by volume.

c. High-gloss finish (minimum 85 GU at a 60-degree angle).

d. Retain gloss and color on prolonged outdoor exposure.

e. PCB-free, lead-free, cadmium-free, and chromate-free (less than 0.002 percent by weight of mixed coating).

f. Listed as an approved coating system (see Appendix B).

g. Attain a numerical rating of not less than 8 in accordance with ASTM D610 and ASTM D1654 and a numerical rating of not less than 9F in accordance with ASTM D714 when applied as a system to composite carbon steel test panels and exposed at the KSC Beach Corrosion Test Site for the following periods:

(1) 18 months for initial acceptance.
(2) 5 years for final acceptance.

4.1.2.4 Epoxy Mastic Coatings

[PCR 40] Epoxy mastic coatings shall conform to the following minimum requirements:

a. Specifically intended for use over mechanically cleaned steel.

b. Contain a minimum of 80 percent solids by volume.

c. Two-component, catalyst-cured, aluminum-pigmented.

d. PCB-free, lead-free, cadmium-free, and chromate-free (less than 0.002 percent by weight of mixed coating).

Examples of epoxy mastic coating that currently meet these requirements include:

(1) Ameron Amerlock 400 AL.
(2) Devoe Bar Rust 239.
(3) PPG Pittguard DTR.
(4) Sherwin-Williams Epolon Mastic.

4.1.2.5 Coal Tar Epoxy

a. [PCR 41] Coal tar epoxy coating shall be a two-component, high-build epoxy of low volatile organic content (VOC).

b. [PCR 42] The coal tar epoxy shall contain, at a minimum, 65 percent solids by volume.

c. [PCR 43] The coal tar epoxy shall produce a one-coat thickness of 405 µm to 510 µm (16 mil to 20 mil) per coat dry film thickness (DFT).

Examples of coal tar epoxies that currently meet these requirements include the following:

a. Sherwin-Williams Hi-Mil Sher-Tar.
b. Made Well 1103.
c. Devoe Devtar 247.

4.1.2.6 Potable Water Epoxy

[PCR 44] All coatings for potable water immersion service shall be three-coat epoxy systems that are certified by NSF Standard 61, Drinking Water System Components – Health Effects.

Some NSF-approved products include:

a. Ameron Amercoat 395.
b. Devoe Bar Rust 233.
c. Sherwin-Williams Potable Water Epoxy.

4.1.2.7 Nonskid Coating

[PCR 45] Approved nonskid coatings shall conform to MIL-PRF-24667, Coating System, Non-skid, for Roll, Spray, or Self-Adhering Application, Type 1, Composition G, as supplied by American Safety Technologies, Inc., 565 Eagle Rock Avenue, Roseland, NJ 07068, telephone (800) 631-7841 (< www.astantislip.com >), or an approved equivalent (Primer MS 7C, Topcoat MS 400G, Color Topping MS-200).

4.1.3 Sealants/Caulking

a. [PCR 46] Sealants shall be self-curing, single-component, polysulfide rubber or polyurethane material only, conforming to ASTM C920, Standard Specification for Elastomeric Joint Sealants, Type S, Grade NS, Class 25, use NT, A, and O.

b. [PCR 47] If not topcoated, the caulking shall match the color of the joint surface being caulked.

c. [PCR 48] If caulking is to be used in a clean-room environment, an approved material with low offgassing characteristics in accordance with NASA-STD-6001 shall be selected.

4.1.4 Chip-Free Clean-Room Paint

[PCR 49] Paint systems for metal substrates in clean-rooms may be required to pass tests for adhesion, offgassing, flammability, vacuum outgassing, and hypergolic compatibility. Offgassing, flammability, and hypergolic compatibility testing shall be in accordance with NASA-STD-6001, Supplemental Test Procedure A.7.

4.2 Equipment

4.2.1 Compressed Air

a. [PCR 50] The compressed air system shall be capable of delivering a continuous nozzle pressure to achieve the required surface cleanliness and profile, typically 620 kPa (90 psi) minimum to each blast nozzle in operation.

The required air capacity depends upon the configuration of the abrasive system used. The air system should comply with the instructions and recommendations of the manufacturer of the abrasive-blasting system.

b. [PCR 51] The compressed air system shall be equipped with oil and moisture separators to ensure only clean, dry air is provided to the service outlet.

c. [PCR 52] The compressed air system shall comply with Occupational Safety and Health Administration (OSHA), American National Standards Institute (ANSI), and National Institute of Occupational Safety and Health (NIOSH) configurations.

d. [PCR 53] Air distribution manifolds shall conform to American Society of Mechanical Engineers (ASME) standards.

4.2.2 Abrasive-Blasting System

a. [PCR 54] The abrasive-blasting system shall comply with OSHA, ANSI, and NIOSH configurations consisting of, but not limited to, the following:

(1) A remote-controlled welded pressure pot conforming to ASME standards.
(2) The required length of blast hose.
(3) A venturi nozzle.
(4) A respiratory air-line filter.
(5) A blast hood approved by the Mine Safety and Health Administration/NIOSH with the required length of air hose.

b. [PCR 55] The blasting system shall be designed to produce the specified cleanliness level and profile when coupled with the available compressed air supply.

4.2.3 Coating Application System

[PCR 56] The coating application equipment shall be an airless spray system, conventional spray system, or other approved equipment in accordance with the coating manufacturer's recommendations and section 4.4.3.6.

4.2.4 Respiratory Protection

[PCR 57] Respiratory protection shall be in accordance with 29 CFR 1910.134, Respiratory Protection, and Center respiratory protection requirements.

4.3 Safety and Health Requirements

a. [PCR 58] Necessary precautions, in accordance with OSHA regulations, manufacturers' recommendations, and industry standards, shall be taken to ensure the safety and health of personnel performing the work required by this NASA Technical Standard and personnel who may be affected by such work.

Some of the materials handled in accordance with this document are combustible, or toxic, or both.

b. [PCR 59] The Contractor shall provide equipment as required for safe and healthful application and instruction to the users regarding the hazards and proper handling and disposal procedures to prevent injury or illness.

c. [PCR 60] The Contractor shall provide safe access to all areas for the coating inspector.

d. [PCR 61] The Contractor shall submit a written safety and health plan that includes a Hazard Communication Program, a Respiratory Protection Program, and a Hearing Conservation Program that conforms to OSHA requirements and industry standards.

e. [PCR 62] Where the contractor is required to remove surface coatings that contain PCB, lead, chromium, mercury, or cadmium, or other regulated materials, the Contractor shall include specific provisions in the safety and health plan for complying with all Federal, State, Local, and NASA Center-specific requirements.

4.3.1 Environmental Requirements

The operations described in this NASA Technical Standard have the potential to impact the environment.

a. [PCR 63] All local, state, and federal environmental regulations, as well as the NASA Center's environmental policies, shall be followed.

b. [PCR 64] Questions regarding these regulations and policies shall be directed to the local environmental management organization.

4.3.2 Personal Protective Equipment (PPE)

a. [PCR 65] When engineering controls cannot be implemented to protect workers, then PPE and/or administrative controls shall be used.

b. [PCR 66] Where required, PPE shall be used in accordance with all federal, state, NASA, and Center requirements.

c. [PCR 67] Both the supervisors and the workers shall be properly instructed, trained, and certified in the selection, use, and maintenance of PPE.

4.4 General Requirements

4.4.1 Applicator Qualifications

a. [PCR 68] To ensure the highest quality of workmanship, only coating applicators who have worked in the painting trade sufficiently long enough to master the use of all applicable tools and materials shall be assigned to perform the work described herein.

The applicator's proficiency and ability to attain the required quality of workmanship for the specified coating system can be verified by testing and qualification in accordance with ASTM D4228, Standard Practice for Qualification of Coating Applicators for Application of Coatings to Steel Surfaces.

b. [PCR 69] In addition, the coating applicators shall provide written evidence of having successfully completed a comprehensive training program, such as Painting and Decorating Contractors of America (PDCA)/NACE/SSPC Industrial Painters Training, or equivalent.

c. [PCR 70] The Contractor shall provide all painting personnel an orientation on the proper mixing and application of the coatings specified, particularly inorganic zinc coatings.

d. [PCR 71] Topics in the orientation of proper mixing and application of the specified coatings (particularly for inorganic zinc coatings) shall include specification requirements, material application characteristics, and inspection criteria.

e. [PCR 72] The mixing or application of coatings shall be performed only by personnel who have received training.

f. [PCR 73] The Contractor shall prepare representative sample areas that meet specification requirements.

4.4.2 Preparation of Surfaces

a. [PCR 74] All surfaces to be coated shall be clean, dry, and free from oil, grease, dirt, dust, corrosion, peeling paint, caulking, weld spatter, and any other surface contaminants.

b. [PCR 75] All surfaces that cannot be accessed after fabrication, erection, or installation shall be prepared and coated while accessible.

c. [PCR 76] Surface preparation and coating operations shall be sequenced, so that freshly applied coatings will not be contaminated by dust or foreign matter.

d. [PCR 77] All equipment and adjacent surfaces not to be coated shall be protected from surface preparation operations.

e. [PCR 78] Working mechanisms shall be protected against intrusion of the abrasive.

f. [PCR 79] All surfaces shall be degreased, as required, before subsequent surface preparation procedures or the application of protective coatings, or both.

g. [PCR 80] The following sections provide the surface preparation techniques that shall be used when specified in section 4.5.

4.4.2.1 Cleaning and Degreasing

a. [PCR 81] Degreasing shall be by solvent cleaning, detergent washing, or steam cleaning in accordance with SSPC-SP 1, Solvent Cleaning.

b. [PCR 82] This degreasing procedure shall be followed when cleaning carbon steel, galvanized steel, stainless steel, or aluminum.

c. [PCR 83] Selection of solvents shall be in accordance with use requirements and applicable federal, state, and NASA environmental policies.

d. [PCR 84] Chlorofluorocarbon solvents shall not be used.

e. [PCR 85] Water washing, using clean potable water, shall be done when high levels of chloride ($>5\ \mu g/cm^2$) or other undesirable contaminants are found on the surfaces.

f. [PCR 86] Water washing shall be accomplished using standard industrial pressure cleaners with a pressure-versus-volume output balance that will ensure thorough and productive cleaning.

g. [PCR 87] All water washing or pressure cleaning operations shall comply with all Federal, State, Local, and NASA Center environmental requirements.

h. [PCR 88] The cleaned surface shall be free of loose coatings, chlorides, dirt, dust, mildew, grinding/welding/cutting debris, and visible contaminants.

i. [PCR 89] The surface shall be clean and dry prior to the abrasive-blasting operations and application of coatings.

4.4.2.2 Abrasive Blasting

a. [PCR 90] The abrasive-blasting aggregate shall be clean and dry and conform to section 4.1.1.

b. [PCR 91] The abrasive-blasting system shall conform to section 4.2.2.

c. [PCR 92] Abrasive blasting shall be in accordance with the applicable paragraphs in section 4.5.

d. [PCR 93] Abrasive residues shall be removed from the surface, leaving it clean and dry before the coatings are applied.

e. [PCR 94] All particulate emissions generated during abrasive-blasting operations shall be contained.

f. [PCR 95] The containment system shall be designed to comply with all applicable federal, state, and local regulations as well as all NASA policies.

g. [PCR 96] Exemptions to the requirement in 4.4.2.2f shall be coordinated with the local environmental management office.

h. [PCR 97] The aggregate used to prepare abrasive-sensitive hardware such as bellows, gimbal joints, and other thin-walled components shall be carefully identified and selected.

4.4.2.3 Mechanical Cleaning Methods

[PCR 98] Mechanical methods shall be in accordance with the applicable paragraph in section 4.5.

4.4.3 Application of Coatings

a. [PCR 99] All prepared surfaces shall be coated within 6 hours after surface preparation and before corrosion or recontamination occurs. *As an exception, surfaces prepared under temperature and humidity control may be coated after 6 hours but only after inspection of the surface preparation confirms that the cleanliness level has met the specified standards.*

b. [PCR 100] Any surface that shows corrosion or contamination, regardless of the length of time after preparation, shall be prepared again.

c. [PCR 101] Because the application and handling characteristics of all coatings will vary, adequate written instructions from the manufacturer are essential and shall be closely followed in conjunction with the requirements defined herein to obtain optimum performance.

d. [PCR 102] The manufacturer's written recommendations for thinning, mixing, handling, and applying the product shall be strictly followed.

e. [PCR 103] All coatings shall be thoroughly worked into all joints, crevices, and open spaces.

f. [PCR 104] All newly coated surfaces shall be protected from damage.

g. [PCR 105] All equipment and adjacent surfaces not to be coated shall be protected from overspray and splattered coatings.

h. [PCR106] Particulate emissions shall be contained during all spray-painting operations.

i. [PCR 107] The containment system shall be designed to comply with all federal, state, and local regulations as well as all NASA policies.

j. [PCR 108] Exemptions to this requirement shall be coordinated with the local environmental management organization.

4.4.3.1 Coating Systems

a. [PCR 109] Coating systems for specified uses and substrates shall be as defined in section 4.5 and conform to section 4.1.2.

b. [PCR 110] All thinners and cleaners shall be products of the coating manufacturer, except as defined in section 4.1.2.7.

c. [PCR 111] To ensure intercoat compatibility, coating systems consisting of more than one coat shall be products of the same manufacturer.

d. [PCR 112] Continuity of the coating manufacturer's system shall be maintained for the duration of an individual project.

4.4.3.2 Colors

a. [PCR 113] Inorganic zinc coatings shall be pigmented so that there is a definite contrast between the coating and the dull gray appearance of the blasted steel surface during the coating application.

b. [PCR 114] Color coding for fluid system piping shall be in accordance with KSC-STD-SF-0004, Ground Piping Systems Color Coding and Identification, Safety Standard for.

c. [PCR 115] Finish coat colors shall be in accordance with the following FED-STD-595, Colors Used in Government Procurement, color numbers using pigments free of PCB, lead, chromium, and cadmium:

(1) White, No. 17925.
(2) Blue, No. 15102 (safety).
(3) Yellow, No. 13538 (standard).
(4) Yellow, No. 13655 (safety).
(5) Red, No. 11136.
(6) Red, No. 11105 (safety).
(7) Black, No. 17038.
(8) Green, No. 14110 (safety).
(9) Gray, No. 16187 (safety).
(10) Brown, No. 10080 (safety).
(11) Gray, No. 16473 (standard).

4.4.3.3 Storage of Coating Materials

a. [PCR 116] Coating materials and thinners shall be stored in their original containers with the manufacturer's name, product identification, shelf life, and batch number.

b. [PCR 117] Coating materials, thinners, and cleaners shall be stored in tightly closed containers in a covered, well-ventilated area where they will not be exposed to sparks, flame, direct sunlight, extreme heat, or rainfall.

c. [PCR 118] The manufacturer's written instructions for storage limitations shall be followed.

d. [PCR 119] Tarpaulins shall not be used as the sole means of covering coating materials for storage.

e. [PCR 120] Material Safety Data Sheets for coating materials and thinners shall be maintained or made accessible to users in the area.

f. [PCR 121] The Contractor shall submit a written plan for approval for storage of coating materials for coordination with the local safety/fire/environmental organization.

4.4.3.4 Mixing and Application Instructions

a. [PCR 122] Coating materials shall be thoroughly mixed prior to application with a mechanical mixing instrument that will not induce air into the coating, such as a Jiffy Mixer, manufactured by the Jiffy Mixer Company (< www.jiffymixer.com >), Inc., Riverside, CA, or an approved equivalent.

b. [PCR 123] The mixer shall be powered by an air motor or an explosion-proof electric motor.

c. [PCR 124] All mixing operations shall be performed over an impervious surface with provisions to prevent runoff to grade of any spilled material.

d. [PCR 125] The mixed coating material shall be strained through a 30-mesh to 60-mesh screen prior to application.

e. [PCR 126] Thinning shall be for viscosity control only.

f. [PCR 127] If thinner is required, the amount recommended by the manufacturer of the thinner shall be used.

g. [PCR 128] The material shall be agitated as required during application to maintain a uniform suspension of solids.

h. [PCR 129] Continuous rapid agitation shall be avoided.

i. [PCR 130] Spray equipment shall be adjusted to produce an even, wet coat with minimum overspray.

j. [PCR 131] The conventional pressure pot, when used, shall be kept at approximately the same level or above the spray gun, so that the material is delivered properly.

k. [PCR 132] Coatings shall be applied in even, parallel passes, overlapping 50 percent.

4.4.3.5 Weather Conditions

a. [PCR 133] No coating shall be applied when contamination from any source (i.e., rainfall) is imminent or when the temperature or humidity is outside limits recommended by the coating manufacturer.

b. [PCR 134] To prevent condensation during application, the surface temperature shall be at least 3 °C (5 °F) above the dew point and rising.

c. [PCR 135] Spray application methods shall not be used when wind speed exceeds 25 km/hr (15 mph) in the area where the coating is being applied.

d. Limitations against using certain coatings under specific relative humidity (RH) are as follows:

(1) [PCR 136] Solvent-based inorganic zinc coatings, polysiloxane topcoats, and IOTs shall not be applied in conditions with <40 percent RH.

(2) [PCR 137] Water-based inorganic zinc coatings shall not be applied in conditions with <40 percent or >80 percent RH.

4.4.3.6 Methods of Application

a. [PCR 138] Coatings shall be applied with airless or conventional spray equipment, or both, according to section 4.2.3.

b. [PCR 139] Application with brushes shall be permitted for minor touchup of spray applications and stripe coats of inorganic zinc.

Organic midcoats and topcoats may be applied using a brush, roller, or spray device as applicable.

4.4.3.7 Coating Finish

a. [PCR 140] Each coat of material applied shall be free of runs, sags, blisters, bubbles, and mud-cracking; variations in color, gloss, and texture; holidays (missed areas); excessive film buildup; foreign contaminants; dry overspray; etc.

b. [PCR 141] Special care shall be taken to ensure complete coverage and proper thickness on welds, corners, crevices, sharp edges, bolts, nuts, and rivets.

c. [PCR 142] Each coat of applied material shall be rendered clean, dry, and free from surface contaminants before another coating is applied.

4.4.3.8 Touchup of Welds and Damaged Coatings

a. [PCR 143] Field welds and damaged coatings shall be touched up in accordance with section 4.5.8.

b. [PCR 144] The coating shall be applied in accordance with sections 4.4.3.4 and 4.4.3.6.

c. [PCR 145] Touchup and repair shall be accomplished promptly after the damage or welding has occurred.

4.4.3.9 Coating, Drying, and Curing

a. [PCR 146] The coating manufacturer's recommended drying and curing times for handling, recoating, and topcoating shall be followed.

b. [PCR 147] The coating manufacturer's recommendations shall be followed to test the coating for proper curing.

c. [PCR 148] Proper curing of solvent-based inorganic zinc-rich coatings shall be verified by ASTM D4752, Standard Practice for Measuring MEK Resistance of Ethyl Silicate (Inorganic) Zinc-Rich Primers by Solvent Rub, prior to further coating.

d. [PCR 149] Water-based inorganic zinc-rich coatings shall be verified for curing in accordance with the same procedure, but water is to be substituted as the solvent.

Note: The curing time of solvent-based inorganic zinc coatings can be accelerated by rinsing or spraying the coating with potable fresh water after an initial overnight drying. The number and frequency of rinse cycles can vary with environmental conditions. Check with the material manufacturer for recommended procedures.

4.4.4 Sealing/Caulking

a. [PCR 150] The perimeter of all faying surfaces, joints open less than 13 mm (0.5 in), and skip-welded joints shall be completely sealed.

b. [PCR 151] The sealant shall be a self-curing, single-component, polysulfide rubber or polyurethane type, conforming to section 4.1.3.

c. [PCR 152] The sealant shall be applied to the joint with a caulking gun after the inorganic zinc primer has been applied on carbon steel.

d. [PCR 153] For topcoated zinc primers, caulking shall be applied after the intermediate coat of epoxy.

e. [PCR 154] For coatings on stainless steel, galvanized steel, and aluminum, caulking shall be applied before application of the topcoat.

f. [PCR 155] The bead shall have a smooth and uniform finish and be cured (tacky to the touch) before the topcoat is applied.

4.5 Specific Requirements

4.5.1 Protection of Carbon Steel

a. [PCR 156] Carbon steel surfaces shall be protected from atmospheric corrosion through the application of zinc coatings (inorganic zinc coating and/or hot-dip galvanizing and/or metallizing) as defined herein.

b. [PCR 157] New steel components, such as stair treads, grating, handrails, pipes, and hardware (nuts, bolts, and fasteners), shall be hot-dip galvanized in accordance with section 4.5.1.2.1, as applicable.

c. [PCR 158] All other carbon steel surfaces that are exposed to the atmosphere shall be coated with inorganic zinc conforming to section 4.1.2 in accordance with section 4.4.3, hot-dip-galvanized (zinc-coated) in accordance with section 4.5.1.2.1, or metallized in accordance with section 4.5.1.3.

The zinc coatings may require topcoating with additional protective coatings as specified; but in neutral pH atmospheres, testing has proven zinc without topcoating to have superior performance.

d. [PCR 159] Carbon steel faying surfaces that are a part of all friction-type and electrical grounding joints shall be abrasive-blasted and coated with 100 μm to 150 μm (4 mil to 6 mil) of inorganic zinc only, in accordance with section 4.5.1.1.4, prior to installation.

e. [PCR 160] An inorganic zinc coating used in a friction-type joint shall be approved by the American Institute of Steel Construction (AISC).

f. [PCR 161] The recommended coating application sequence for carbon steel shall be to first abrasive blast the steel and then to prime it with inorganic zinc before installation or erection.

g. [PCR 162] Further topcoating, if required, shall be done after all welding, grinding, or drilling has been completed, and after areas damaged by these procedures have been properly repaired with inorganic zinc.

4.5.1.1 Protection with Inorganic Zinc

4.5.1.1.1 Pre-Cleaning of Carbon Steel

[PCR 163] Carbon steel surfaces shall be cleaned and degreased in accordance with SSPC-SP 1 followed by power tool cleaning in accordance with SSPC-SP 3, Power Tool Cleaning, to remove weld spatter, weld slag, laminations, sharp edges, and other surface defects prior to abrasive blasting or power tool cleaning to bare metal.

4.5.1.1.2 Power Tool Cleaning of Carbon Steel

a. [PCR 164] Carbon steel shall be cleaned to bare metal, using power tools, in accordance with SSPC-SP 11, Power Tool Cleaning to Bare Metal, when a roughened, clean, bare metal surface is required but abrasive blasting is not feasible or permissible.

b. [PCR 165] The surface anchor profile of the surface cleaned with the power tool shall be 40 μm to 75 μm (1.5 mil to 3.0 mil).

c. [PCR 166] All rust shall be completely removed from pits and depressions.

4.5.1.1.3 Abrasive Blasting of Carbon Steel

a. [PCR 167] Carbon steel shall be abrasive-blasted to a minimum cleanliness of near-white metal, in accordance with SSPC-SP 10/NACE No. 2, Near-White Metal Blast Cleaning, with aggregate conforming to the requirements in section 4.1.1.

b. [PCR 168] The anchor profile of the blasted surface shall be 40 µm to 75 µm (1.5 mil to 3.0 mil), measured in accordance with ASTM D4417, Standard Test Methods for Field Measurement of Surface Profile of Blast Cleaned Steel.

c. [PCR 169] All rust shall be completely removed from pits and depressions.

4.5.1.1.4 Stripe Coat Application

[PCR 170] Stripe coating with inorganic zinc shall be applied to welds, cutouts, sharp edges, rivets, crevices, and bolts to ensure complete coverage prior to subsequent applications of inorganic zinc.

4.5.1.1.5 Application of Inorganic Zinc Coatings

a. [PCR 171] Inorganic zinc coatings shall be applied to a DFT of 100 µm (4.0 mil) minimum to 150 µm (6.0 mil) maximum when they will be left without a topcoat or when IOT or ablative coating is applied.

b. [PCR 172] When the zinc coatings are to be topcoated with organic topcoats, the DFT shall be reduced to 65 µm (2.5 mil) minimum to 100 µm (4.0 mil) maximum.

c. [PCR 173] The proper DFT for the inorganic zinc coating shall be obtained in a single application, which may consist of multiple passes, while the coating is still wet (including the application of a stripe coat).

4.5.1.1.6 Topcoat Systems for Inorganic Zinc Coatings

a. [PCR 174] The following topcoat systems shall be applied over the inorganic zinc coatings as required for each zone of exposure described in section 1.4.

b. [PCR 175] Topcoats shall be applied at the DFT recommended by the manufacturer or as specified in the section that follows.

c. [PCR 176] The film thickness of the topcoats shall be sufficient to ensure uniform coverage and color.

(1) Zones 1a and 1b. Inorganic zinc coatings may be left without a topcoat; however, for maximum protection, the inorganic zinc coating should be topcoated with a heat-resistant silicone ablative coating material in accordance with KSC-SPEC-F-

0006, Heat and Blast Protection Coating Materials and Application Methods, Specification for.

(2) [PCR 177] Zone 1c. Inorganic zinc coatings shall be left without a topcoat.

(3) [PCR 178] Zone 2. An IOT conforming to section 4.1.2.3.3 shall be applied at a DFT of 75 µm to 125 µm (3 mil to 5 mil).

As an alternate, surfaces may be top coated with a heat-resistant silicone ablative coating material in accordance with KSC-SPEC-F-0006.

(4) Zones 3a and 3b.

A. [PCR 179] An intermediate/tie coat and a finish coat conforming to section 4.1.2 shall be applied in accordance with section 4.4.3.

B. [PCR 180] As an alternate, an IOT conforming to section 4.1.2.3.3 or a polysiloxane finish coat conforming to section 4.1.2.3.4 shall be applied at the manufacturer's recommended DFT.

C. [PCR 181] The DFT shall be sufficient to completely hide the inorganic zinc primer.

(5) Zones 4a, 4b, and 4c. No topcoats are required, except as needed for color coding, safety purposes, identification, or special conditions.

[PCR 182] When required for color coding, safety purposes, identification, or special conditions, topcoats shall be in accordance with section 4.4.3.2.

(6) Zone 5a and 5b. Inorganic zinc coating is suggested but not required. *As an alternate, an inhibitive epoxy primer and a polyurethane finish coat conforming to section 4.1.2 at the manufacturer's recommended thickness may be used.*

(7) [PCR 183] Zone 6. The coating system shall be as specified in sections 4.5.4 and 4.5.5.

(8) [PCR 184] Zone 7. The coating system shall be as specified in NACE International RP0198, Control of Corrosion Under Thermal Insulation and Fireproofing Materials – A Systems Approach.

4.5.1.2 Protection by Galvanizing

4.5.1.2.1 Galvanizing

a. [PCR 185] Galvanizing (zinc coating) shall be accomplished after fabrication by the hot-dip process conforming to ASTM A123, Standard Specification for Zinc (Hot-Dip Galvanized) Coatings on Iron and Steel Products; ASTM A153, Standard Specification for Zinc Coating (Hot-Dip) on Iron and Steel Hardware; and ASTM A653, Standard Specification for Steel Sheet, Zinc-Coated (Galvanized) or Zinc-Iron Alloy-Coated (Galvannealed) by the Hot-Dip Process.

b. [PCR 186] Galvanizing weight for steel sheet without further coating protection shall meet the standards of ASTM A653, with a galvanizing weight of G165.

c. [PCR 187] All lower galvanizing weights for steel sheet shall be further protected with coatings except for Zone 5a and 5b exposures.

d. [PCR 188] In accordance with this NASA Technical Standard, the galvanneal process shall not be used for the coating of steel sheet.

e. [PCR 189] Steel components with an ultimate tensile strength above 900 MPa (130 ksi) or hardness above Rockwell C Hardness 28 shall not be galvanized due to potential hydrogen embrittlement.

4.5.1.2.2 Surface Preparation for Galvanizing

CAUTION: *Some galvanized configurations are susceptible to distortion when they are abrasive-blasted.*

a. [PCR 190] Special care shall be taken to prevent any metal distortion by reducing blast nozzle pressure and increasing the working distance from the nozzle to the surface.

b. [PCR 191] In some cases, such as in the surface preparation of light-gage sheet steel, these precautions may not be sufficient to prevent distortion; and alternate procedures, such as abrading or mechanical cleaning, shall be used to remove corrosion or roughen the surface.

c. [PCR 192] Galvanized surfaces shall be abrasive-blasted with fine-grade abrasives conforming to the requirements in section 4.1.1 to remove corrosion and old coatings or roughen new surfaces.

d. [PCR 193] The blasted surface shall be free of all corrosion and foreign matter and have a uniform, slightly roughened appearance.

e. [PCR 194] Galvanized surfaces to be further topcoated shall be prepared by degreasing in accordance with section 4.4.2.1 before any additional surface preparation.

f. [PCR 195] After degreasing, abrasive blasting or mechanical cleaning shall be performed as required by the zone of exposure, as defined in section 4.5.1.2.3.

g. [PCR 196] If galvanized steel is prepared for the application of coatings by abrasive blasting, it shall be lightly brush-blasted with fine-grade abrasive at a lower pressure to provide a corrosion-free and uniform, slightly roughened surface.

h. [PCR 197] Care shall be taken not to completely remove the galvanized finish.

i. [PCR 198] The zinc coatings shall be maintained or rendered clean, dry, and free from contaminants before topcoat systems are applied.

j. [PCR 199] Field repair of damaged galvanized surfaces shall be accomplished in accordance with ASTM A780, Standard Practice for Repair of Damaged and Uncoated Areas of Hot-Dip Galvanized Coatings, using inorganic zinc coatings.

k. [PCR 200] Galvanized steel that is to be mechanically cleaned shall be cleaned in accordance with SSPC-SP 3 using abrasive discs/sheets, or other approved methods.

l. [PCR 201] All corrosion and foreign matter shall be completely removed and the entire surface slightly roughened.

4.5.1.2.3 Coating Systems for Galvanizing

a. [PCR 202] Zones 1a and 1b. Galvanized surfaces may be left without a topcoat; however, for maximum protection, the galvanized coating shall be topcoated with a heat-resistant silicone ablative coating material in accordance with KSC-SPEC-F-0006.

b. [PCR 203] Zone 1c. Galvanized surfaces shall be left without a topcoat.

c. [PCR 204] Zone 2. After brush-blasting, an IOT conforming to section 4.1.2.3.3 shall be applied at a DFT of 75 μm to 125 μm (3 mil to 5 mil).

As an alternate, surfaces may be topcoated with a heat-resistant coating material, such as a silicone ablative coating material in accordance with KSC-SPEC-F-0006.

d. [PCR 205] Zones 3a and 3b. After brush-blasting, primer/tiecoat and finish coat conforming to section 4.1.2 shall be applied in accordance with manufacturer's recommended thicknesses.

(1) [PCR 206] As an alternate, an IOT conforming to section 4.1.2.3.3 or a polysiloxane finish coat conforming to section 4.1.2.3.4 shall be applied at the manufacturer's recommended DFT.

(2) [PCR 207] The DFT shall be sufficient to completely hide the galvanized coating.

e. Zones 4a, 4b, and 4c. *No topcoats are required for galvanizing weights meeting or exceeding ASTM A123, ASTM A153, and ASTM A653, with a galvanizing weight of G165.*

(1) [PCR 208] When steel sheet is galvanized less than ASTM A653, with a galvanizing weight of G165, further coating in accordance with Zone 3 shall be required.

(2) [PCR 209] As an alternate to topcoats, steel sheet shall be degreased, brush-blasted, and an inorganic zinc primer conforming to section 4.1.2.1 applied to a DFT of 50 μm to 75 μm (2 mil to 3 mil).

f. Zone 5a and 5b. *No topcoats are required, except for when needed for color coding, safety purposes, identification, or special conditions.*

(1) [PCR 210] When topcoats are required for color coding, safety purposes, identification, or special conditions, the surface shall be degreased and an epoxy primer applied at the manufacturer's recommended DFT.

(2) [PCR 211] Within 24 hours, a polyurethane finish coat conforming to section 4.1.2 shall be applied at the manufacturer's recommended DFT.

(3) [PCR 212] As an alternate, a polysiloxane topcoat conforming to section 4.1.2.3.4 shall be applied at the manufacturer's recommended DFT.

(4) [PCR 213] The DFT shall be sufficient to completely hide the galvanized coating.

g. [PCR 214] Zone 6. The coating system shall be as specified in sections 4.5.4 and 4.5.5.

h. [PCR 215] Zone 7. The coating system shall be as specified in NACE International RP0198.

4.5.1.3 Protection with Metallizing

4.5.1.3.1 Pre-Preparation of Carbon Steel

[PCR 216] Carbon steel surfaces shall be cleaned and degreased in accordance with SSPC-SP 1 followed by power tool cleaning in accordance with SSPC-SP 3 to remove weld spatter, weld slag, laminations, sharp edges, and other surface defects prior to abrasive blasting or power-tool cleaning to bare metal.

4.5.1.3.2 Abrasive Blasting of Carbon Steel

a. [PCR 217] At a minimum, carbon steel shall be abrasive-blasted to near-white metal (SSPC-SP 10/NACE No. 2) with aggregate conforming to the requirements in section 4.1.1.

b. [PCR 218] The anchor profile of the blasted surface shall be 62.5 μm to 75 μm (2.5 mil to 3 mil).

c. [PCR 219] All rust shall be completely removed from pits and depressions.

4.5.1.3.3 Application of Metallized Coatings

a. [PCR 220] Metal wire to be used with the arc spray metallizing equipment shall be pure zinc, 90-10 zinc-aluminum, 85-15 zinc-aluminum alloys, 53-56 aluminum-magnesium, or pure magnesium.

b. [PCR 221] Metallized zinc coatings shall be applied to a DFT of 200 μm (8 mil) minimum to 375 μm (15 mil) maximum, depending on the intended service environment.

4.5.1.3.4 Topcoat Systems for Metallized Zinc Coatings

a. [PCR 222] Topcoat systems shall be applied over the metallized zinc coatings as required for each zone of exposure described in section 1.1.

b. [PCR 223] The coating materials shall be selected from Appendix C, Approved Products List for Metallized (TSC) Systems.

c. [PCR 224] Topcoats shall be applied at the DFT recommended by the manufacturer or as specified in requirement "d" that follows.

d. [PCR 225] The film thickness of the topcoats shall be sufficient to ensure uniform coverage and color.

(1) [PCR 226] <u>Zones 1a and 1b.</u> Metallized coatings may be left without a topcoat; however, for maximum protection, the metallized coating shall be topcoated with a heat-resistant silicone ablative coating material in accordance with KSC-SPEC-F-0006.

(2) [PCR 227] <u>Zone 1c.</u> Metallized coatings shall be left without a topcoat.

(3) [PCR 228] <u>Zone 2.</u> An IOT conforming to section 4.1.2.3.3 shall be at a DFT of 75 μm to 125 μm (3 mil to 5 mil).

As an alternate, surfaces may be topcoated with a heat-resistant silicone ablative coating material in accordance with KSC-SPEC-F-0006.

(4) [PCR 229] Zone 3. An intermediate/tie coat and a finish coat conforming to section 4.1.2 shall be applied in accordance with section 4.4.3.

A. [PCR 230] As an alternate, an IOT conforming to section 4.1.2.3 shall be applied at a DFT of 75 µm to 125 µm (3 mil to 5 mil) or a polysiloxane finish coat conforming to section 4.1.2.3.4 applied at the manufacturer's recommended DFT.

B. [PCR 231] The DFT shall be sufficient to completely hide the metallized coating.

(5) Zones 4a, 4b, and 4c. *No topcoats are required, except as needed for color coding, safety purposes, identification, or special conditions.*

[PCR 232] When topcoats are required for color coding, safety purposes, identification, or special conditions, topcoats shall be in accordance with section 4.4.3.2.

(6) Zone 5a and 5b. A *metallized coating is suggested but not required. As an alternate, an inhibitive epoxy primer and a polyurethane finish coat conforming to section 4.1.2 may be applied at the manufacturer's recommended thickness.*

(7) [PCR 233] Zone 6. The coating system shall be as specified in sections 4.5.4 and 4.5.5.

(8) [PCR 234] Zone 7. The coating system shall be as specified in NACE International RP0198.

4.5.2 Protection of Aluminum

[PCR 235] Aluminum shall be protected from corrosion by the use of protective coatings as defined herein. *Certain alloys may require coatings in specific environments as specified in section 4.5.2.2.*

4.5.2.1 Surface Preparation of Aluminum

CAUTION: *Some aluminum configurations are susceptible to distortion and/or destruction when they are abrasive-blasted.*

a. [PCR 236] Special care shall be taken to ensure against any metal damage by the choice of abrasive aggregate and by reducing blast nozzle pressure and increasing the working distance from the nozzle to the surface as necessary.

b. [PCR 237] In some cases, such as in the surface preparation of light-gage sheet, these precautions may not be sufficient to prevent distortion; and an alternate procedure, such as abrading or mechanical cleaning, shall be used to remove corrosion or roughen the surface.

c. [PCR 238] Aluminum surfaces shall be abrasive-blasted with fine-grade abrasive materials conforming to the requirements in section 4.1.1 to remove corrosion and old coatings or roughen new surfaces.

d. [PCR 239] The blasted surface shall be free of all corrosion and foreign matter and have a uniform, slightly roughened appearance.

e. [PCR 240] Aluminum shall be prepared by degreasing and abrasive blasting or mechanical cleaning, as required by the condition and configuration of the surface.

f. [PCR 241] Abrasive blasting shall be used whenever possible using nonmetallic abrasives specified in section 4.1.1.

g. [PCR 242] Mechanical cleaning shall be used only when abrasive blasting is impractical, would damage the structure or component, or is prohibited in the area where the work is being performed.

h. [PCR 243] Aluminum shall be mechanically cleaned in accordance with SSPC-SP 3 using abrasive discs/sheets, or other approved methods.

i. [PCR 244] All corrosion and foreign matter shall be completely removed and the entire surface slightly roughened.

j. [PCR 245] Anodized or chemical-conversion-coated aluminum surfaces shall not be mechanically cleaned.

k. [PCR 246] In accordance with section 4.1.1, plastic media or an approved equivalent shall be used for abrasive blasting of bellows, gimbal joints, and other thin-walled, abrasion-sensitive components.

4.5.2.2 Protective Coatings

Note: *Aluminum surfaces require special coatings if used underwater. See section 4.5.4 for coatings for underwater use.*

[PCR 247] The following protective coatings shall be applied to aluminum surfaces as required for each zone of exposure described in section 1.4:

a. [PCR 248] <u>Zones 1, 2, and 3.</u> The following coatings shall be used to protect aluminum in the launch environment. *To facilitate washdown of SRB residue on critical hardware, an inhibited polyamide epoxy coating and aliphatic polyurethane topcoat may be used as well as other coatings such as polysiloxane, IOTs, and silicone ablative.*

b. [PCR 249] Zones 4 and 5. *No protective coatings are required except as needed for color coding, safety purposes, identification, or special conditions for normal atmospheric service of 1000-, 5000-, and 6000-series alloys.* Aluminum that is located within 3.5 km (2 mi) of the coastline or subject to chemical exposure shall be fully coated according to section 4.5.2.2a.

c. [PCR 250] Zone 6. The coating system shall be as specified in sections 4.5.4 and 4.5.5.

d. [PCR 251] Zone 7. The coating system shall be as specified in NACE International RP0198.

4.5.3 Protection of Stainless Steel

a. [PCR 252] Type 300 series stainless steels shall be protected from corrosion by the use of protective coatings as defined in section 4.5.3.2.

Certain highly alloyed stainless steels, such as AL6XN or 254 SMO, do not require protective coatings for corrosion protection.

Note: *Thin-walled 300-series stainless-steel tubing is subject to pitting corrosion failure in outdoor marine environments.*

b. [PCR 253] For exterior installations, thin-walled 300-series stainless-steel tubing shall be degreased, prepared with a stainless-steel wire wheel or equivalent, and coated in accordance with section 4.5.3.2.

4.5.3.1 Surface Preparation of Stainless Steel

a. [PCR 254] Stainless steel shall be prepared by degreasing in accordance with SSPC-SP 1 and abrasive blasting or mechanical cleaning.

b. [PCR 255] Abrasive blasting shall be used whenever possible, using nonmetallic abrasives specified in 4.1.1.

c. [PCR 256] As an alternative, stainless steel shall be mechanically cleaned in accordance with SSPC-SP 3 using abrasive discs/sanding sheets or other approved methods.

d. [PCR 257] All corrosion and foreign matter shall be completely removed and the entire surface slightly roughened.

CAUTION: *Some stainless steel configurations are susceptible to distortion and/or destruction, when they are abrasive-blasted.*

e. [PCR 258] Special care shall be taken to ensure against any metal damage by choice of abrasive aggregate and by reducing the blast nozzle pressure and increasing the working distance from the nozzle to the surface as necessary.

f. [PCR 259] In some cases, such as in the surface preparation of light-gage sheet, these precautions may not be sufficient to prevent distortion; and an alternate procedure, such as abrading or mechanical cleaning, shall be used to remove corrosion or roughen the surface.

g. [PCR 260] Stainless steel surfaces shall be abrasive-blasted with fine-grade abrasive conforming to the requirements in section 4.1.1 to remove corrosion and old coatings or roughen new surfaces.

h. [PCR 261] The blasted surface shall be free of all corrosion and foreign matter and have a uniform, slightly roughened appearance.

4.5.3.2 Protective Coating

a. [PCR 262] Zones 1, 2, and 3. For 300 series stainless steels, an inhibited polyamide epoxy primer and aliphatic polyurethane topcoat shall be used.

Other coatings such as polysiloxane and silicone ablative may be substituted as topcoats.

b. [PCR 263] Zones 4 and 5. For special conditions, stainless steel shall be brush-blasted and coated with inhibitive epoxy primer to a DFT of 50 µm to 75 µm (2 mil to 3 mil) followed by a finish coat that provides a DFT of 50 µm to 75 µm (2 mil to 3 mil).

c. [PCR 264] Zone 6. The coating system shall be as specified in sections 4.5.4 and 4.5.5.

d. [PCR 265] Zone 7. The coating system shall be as specified in NACE International RP0198.

4.5.4 Underground, Submerged, or Continuously Wetted Surfaces

a. [PCR 266] Surfaces that will be underground, submerged, or continuously wetted shall be prepared in accordance with SSPC-SP 5/NACE No.1, White Metal Blast Cleaning, with a profile of 75 µm to 100 µm (3 mil to 4 mil) and coated with coal tar epoxy conforming to section 4.1.2.5.

b. [PCR 267] Coal tar epoxy coatings shall not be used for surfaces that will be in contact with potable water.

c. [PCR 268] The coating shall be applied to a minimum DFT of 410 μm (16.0 mil) and checked for missed areas or pinholes with a properly calibrated holiday detector in accordance with NACE International SP0188, Discontinuity (Holiday) Testing of New Protective Coatings on Conductive Substrates.

d. [PCR 269] Cathodic protection requirements shall be coordinated with the application of this coating.

4.5.5 Coating Systems for Potable Water Immersion Service

a. [PCR 270] All surface preparation for carbon steel shall be in accordance with SSPC-SP 5/NACE No. 1, with a surface profile of 75 μm to 100 μm (3 mil to 4 mil).

b. [PCR 271] All coatings for potable water service shall be selected from section 4.1.2.6.

c. [PCR 272] All potable water coating systems shall be inspected in accordance with standard recommended practices in NACE International RP0288, Inspection of Linings on Steel and Concrete, and with SP0188.

4.5.6 Provision for Nonskid Surfaces

[PCR 273] Where a nonskid surface is required for walkways, decks, or other such surfaces, a nonskid coating conforming to section 4.1.2.7 shall be applied as follows:

a. [PCR 274] Carbon steel. Coatings shall be applied directly over the zinc coating (inorganic zinc, galvanizing, or metallizing) and follow surface preparation instructions defined for topcoating in section 4.5.1.

b. [PCR 275] Aluminum and stainless steel. Coatings shall be applied directly over these surfaces after surface preparation following instructions defined for topcoating in sections 4.5.2 and 4.5.3.

4.5.7 Coating Systems for Metallic Surfaces Under Thermal Insulation

[PCR 276] Coating systems for carbon steel and stainless steel surfaces under thermal insulation and cementitious fireproofing shall be as specified in NACE International RP0198.

4.5.8 Repair of Applied Coatings

a. [PCR 277] Newly applied coatings shall be repaired in accordance with table 1, Repair of Applied Coatings.

Table 1—Repair of Applied Coatings

Existing Coating	Repair Coating
Inorganic zinc	
Zones[1] 1 and 4	Inorganic zinc/epoxy mastic for small area touchup
Zone 2	Inorganic zinc/inorganic topcoat
Zones 3 and 5	Epoxy mastic/polyurethane/polysiloxane system for small area touchup
Galvanized steel	
Zones 1 and 4	Inorganic zinc/epoxy mastic for small area touchup
Zone 2	Inorganic zinc/inorganic topcoat
Zones 3 and 5	Epoxy mastic/polyurethane/polysiloxane system for small area touchup
Inorganic topcoat	
All zones	Inorganic zinc/inorganic topcoat
Epoxy/Polyurethane[2]	
Zones 3, 4, and 5	Epoxy/polyurethane system/polysiloxane
Water-reducible	
Zones 3, 4, and 5	Water-reducible intermediate/finish
Coal tar epoxy	
Zone 6	Coal tar epoxy

[1] Zones are defined in section 1.4.
[2] When this coating is replaced with inorganic zinc, complete removal of the existing coating is required.

b. [PCR 278] Surfaces shall be prepared by washing with water and using mechanical methods in accordance with SSPC-SP 11, Power-Tool Cleaning to Bare Metal, to remove corrosion, weld slag, and to "feather back" coating edges.

c. [PCR 279] Touchup and repair shall be accomplished promptly after the damage has occurred.

d. [PCR 280] Touchup and repair of shop-applied coatings shall be accomplished using coatings from the same manufacturer as those applied in the shop.

4.5.9 Maintenance of Existing Coatings

a. [PCR 281] Each support contractor responsible for maintaining facilities or GSE shall develop a Coating Maintenance Plan that includes the following key elements:

(1) Record keeping.

(2) Routine inspection of facilities.

(3) Coating repair criteria.

(4) Coating systems.

(5) Equipment requirements.

(6) Procedures.

(7) Training and certification.

(8) In-process inspection.

(9) Worker protection.

(10) Environmental compliance.

b. [PCR 282] All operations shall be in strict accordance with section 4.3.3.

5. QUALITY ASSURANCE PROVISIONS

5.1 Responsibility for Inspection

5.1.1 [PCR 283] The coating contractor/applicator shall:

a. Provide continuous quality control of all work to ensure complete conformance to the project specifications as defined in section 5.2.

b. Submit a project-specific quality control coating inspection plan to the Contracting Officer for approval.

c. Provide the NASA assigned coatings inspector with safe access to the work.

5.1.2 [PCR 284] The NASA-assigned coatings inspector shall be a NACE Certified Level III inspector under the NACE International Coating Inspector Program (CIP). (*The CIP is provided by NACE International, Education Department, 1440 South Creek Drive, Houston, TX 77084-4906, < www.nace.org >.)*

5.1.3 [PCR 285] Inspection of the surface preparation and coating application processes shall be performed by the NASA-assigned coatings inspector as follows:

a. Perform all of the in-process inspections required by this NASA Technical Standard and the project specifications.

b. Witness, inspect, and test all protective coating work to verify complete compliance with the specified requirements.

c. Document the work on the inspection forms described in section 5.4.

d. Prepare and sign the daily inspection reports on a daily basis and submit them to the Contracting Officer on a weekly basis as a minimum.

e. When a nonconformance report is required, sign and submit it to the Contracting Officer within 1 workday from the time that it is written.

f. After determining that all nonconformances have been corrected and/or the coating work is in compliance with this NASA Technical Standard and the project specifications, complete a conformance verification report for the specific item, area, or project.

g. Sign and seal the conformance verification report. *The application of the certified inspector's seal to the verification conformance report indicates that the inspector personally inspected the indicated work and has found it to be in compliance with the specified requirements.*

h. Not affix the seal to the daily inspection report or to the nonconformance report.

5.2 Requirements for Inspection

a. [PCR 286] <u>Zones 1, 2, and 3.</u> Since these zones are located in the highly corrosive launch environment or other chemical exposures, NACE inspection shall be required for all surface preparation and coating applications, including all new work, touchup of new work, major refurbishment of existing coatings, and modifications.

b. [PCR 287] <u>Zone 4.</u> For systems requiring abrasive blasting and coating of metallic substrates, all surfaces shall require full NACE inspection with the following exception: For touchup of existing coatings, NACE inspection is not mandatory but recommended in cases of critical systems or equipment.

c. [PCR 288] <u>Zones 5a and 5b.</u> All clean-room structures fabricated of aluminum or carbon steel that will be abrasive-blast-cleaned and/or coated outside Zone 5 environments shall require NACE inspection.

All other aluminum or carbon steel structures in Zone 5a environments are exempt from NACE inspection. NACE inspections are required for Zone 5b locations.

d. [PCR 289] Zone 6. Since this zone is located in a highly corrosive underground environment or other submerged exposures, NACE inspection shall be required for all surface preparation and coating applications, including all new work, touchup of new work, major refurbishment of existing coatings, and modifications.

e. [PCR 290] Zone 7. Since this zone is located in a highly corrosive environment, NACE inspection shall be required for all surface preparation and coating applications, including all new work, touchup of new work, major refurbishment of existing coatings, and modifications.

5.3 Inspection Hold Points

[PCR 291] Mandatory inspection hold points shall include, but not be limited to, the following:

a. Verification of ambient weather conditions in accordance with section 4.4.3.5.

b. Prior to beginning of surface preparation work, to include the operation of equipment.

c. After surface preparation work and before the beginning of the coating application work, to include the mixing of products.

d. Before and after the application of each coat of material.

e. After completion and prior to final acceptance.

5.4 Inspection Forms

[PCR 292] All inspections shall be recorded and documented on forms acceptable to the customer.

See Appendices F and G for examples of these forms.

5.5 Inspection Prior to Surface Preparation and Coating Application

[PCR 293] The conditions in the following sections shall be inspected before beginning surface preparation and coating application operations.

5.5.1 Surface Condition

a. [PCR 294] The surface condition shall be visually inspected for compliance with section 4.4.2.

b. [PCR 295] Special attention shall be given to weld spatter, sharp edges, flame or saw cuts, delaminations, burrs, slag, or other surface irregularities that affect performance of protective coatings prior to surface preparation.

5.5.2 Protection of Adjacent Surfaces

a. [PCR 296] Adjacent surfaces shall be visually inspected for adequate protection in accordance with section 4.4.2.

b. [PCR 297] This inspection shall be jointly conducted with a Government Quality Engineering representative.

5.5.3 Ambient Weather Conditions

a. [PCR 298] The ambient weather conditions at the actual location of the work shall be determined before and during the surface preparation and coating application operations to ensure they are correct for the work being conducted.

b. [PCR 299] All measurement instrumentation shall be calibrated per the manufacturer's instructions prior to use.

c. [PCR 300] Proper instrumentation shall be used to measure air temperature, relative humidity, dewpoint, surface temperature, and wind speed and direction.

d. [PCR 301] No spray painting shall proceed when the measured wind speed in the immediate area of the coating work is above 25 km/hr (15 mph).

e. [PCR 302] All of these ambient weather conditions shall be recorded on the Coating System Daily Inspection Report as shown in Appendix F, Coating System Daily Inspection Report.

5.5.4 Compressed Air Cleanliness

a. [PCR 303] The compressed air supply shall be inspected for the use of inline moisture and oil traps.

b. [PCR 304] Proper functioning of the traps shall be evaluated daily by allowing the air supply (down line from the traps) to blow against a clean, white cloth for several minutes, in accordance with ASTM D4285, Standard Test Method for Indicating Oil or Water in Compressed Air.

5.5.5 Surface Salt Concentration

a. [PCR 305] The surface chloride concentration shall be determined on all structures prior to surface preparation operations using an industry-recognized method, such as described in SSPC-TU 4, Field Methods for Retrieval and Analysis of Soluble Salts on Substrates, and recorded in the inspection records weekly.

b. [PCR 306] Surfaces that measure 5 μg/cm^2 (0.00016 oz/ft^2) or above shall require washing with water in accordance with section 4.4.2.1 prior to surface preparation.

5.6 Surface Preparation Inspection

[PCR 307] The inspections in the following sections shall be made to ensure compliance with the surface preparation requirements in section 4.4.2.

5.6.1 Abrasive-Blasting Material

[PCR 308] The abrasive-blasting material shall be verified for compliance with section 4.1.1.

5.6.2 Blast Nozzle Air Pressure and Size

a. [PCR 309] The air pressure at the blast nozzle shall be determined through the use of a hypodermic needle air pressure gage.

b. [PCR 310] The needle of the gage shall be inserted as close to the nozzle as practically possible and in the direction of the air flow.

c. [PCR 311] Pressure readings shall be taken with the blasting system in full operation.

d. [PCR 312] The nozzle pressure shall be recorded.

e. [PCR 313] To ensure the compressor output correlates with the nozzle size, the nozzle shall be checked with a blast nozzle orifice gage initially and then at a frequency determined by the NACE inspector.

5.6.3 Degree of Surface Cleanliness

a. [PCR 314] The surface cleanliness shall be inspected after the surface preparation and before primer application to determine compliance with the applicable requirements of section 4.5.

b. [PCR 315] The degree of cleanliness of abrasive-blasted carbon steel shall be verified with a visual inspection in accordance with section 4.5.1.1.2.

c. [PCR 316] Galvanized steel, aluminum, and stainless steel shall be inspected for cleanliness in accordance with sections 4.5.1.2, 4.5.2, and 4.5.3.

d. [PCR 317] The surface preparation cleanliness requirements defined in section 4.5 shall be applicable to 100 percent of the subject area, including places that are difficult to reach.

Use of SSPC-VIS 1-89, Visual Standard for Abrasive Blast Cleaned Steel, and SSPC-VIS 3, Visual Standard for Power- and Hand-Tool Cleaned Steel, is recommended for judging surface cleanliness.

5.6.4 Surface Profile or Roughness

a. [PCR 318] The anchor profile of an abrasive-blasted carbon steel surface shall be determined by using a surface profile gage, comparator, or replica tape.

b. [PCR 319] The profile shall be in accordance with section 4.5.1.1.2.

c. [PCR 320] Galvanized steel, stainless steel, and aluminum surfaces shall be visually inspected as required for slight roughening in accordance with sections 4.5.1.2, 4.5.2, and 4.5.3.

5.6.5 Blasting of Abrasive-Sensitive Components

a. [PCR 321] Thin-walled, abrasive-sensitive components, such as bellows assemblies or tubing, shall be protected during normal blasting operations in accordance with section 5.5.2.

b. [PCR 322] Walnut shells or an approved equivalent shall be used for surface preparation of these sensitive components in accordance with section 4.1.1 or mechanical methods in accordance with section 4.4.2.3.

5.7 Coating Application Inspection

[PCR 323] The inspections in the following sections shall be made to ensure compliance with the coating application requirements defined in section 4.4.3.

5.7.1 Surface Condition

a. [PCR 324] The prepared surface shall be visually inspected.

b. [PCR 325] The time before coating shall be monitored for compliance with section 4.4.3 before coatings are applied.

5.7.2 Coating Materials

[PCR 326] The coating materials shall be visually inspected for compliance with section 4.4.3.1.

5.7.3 Storage of Coating Material

[PCR 327] Coating material storage conditions shall be periodically inspected for compliance with section 4.4.3.3.

5.7.4 Mixing and Application of Coatings

[PCR 328] The mixing and application of all coatings shall be visually inspected to ensure compliance with sections 4.4.3.4, 4.4.3.6, and 4.4.3.9.

5.7.5 Coating Finish and DFT

a. [PCR 329] The finish and DFT of each applied coating shall be inspected for compliance with sections 4.4.3.7 and 4.5 prior to the application of successive coats.

b. [PCR 330] The DFT measurement on carbon steel shall be taken using a magnetic gage calibrated in accordance with SSPC-PA 2, Procedure for Determining Conformance to Dry Coating Thickness Requirements.

c. [PCR 331] DFT measurements on aluminum and stainless steel shall be taken using an eddy current instrument that has been properly calibrated on surfaces similar to the coated surface.

5.8 Caulking Inspection

[PCR 332] All surfaces shall be visually inspected to determine whether they comply with the requirements for sealing and caulking in accordance with section 4.4.4.

5.9 Galvanizing Inspection

[PCR 333] Galvanized carbon steel shall be inspected in accordance with the applicable ASTM standard in section 4.5.1.2.1.

APPENDIX A

APPROVED PRODUCTS LIST FOR INORGANIC ZINC COATINGS

This list shall be used by or for the Government in the procurement of products covered by this NASA Technical Standard, and such listing of a product is not intended to and does not connote endorsement of the product by NASA. All products listed herein have been tested and meet the requirements for the product as specified. This list is subject to change without notice; revisions or amendments of this list will be issued as necessary. The listing of a product does not release the supplier from compliance with the specification requirements. This list is arranged in two sections based on the coating material's VOC. Use of the information shown herein for advertising or publicity purposes is strictly forbidden.

Thinners and cleaners for each of these coatings shall be procured from the manufacturer of the coating in accordance with sections 4.4.3.1 and 4.4.3.4.

The Materials Test and Corrosion Control Branch in the Engineering Directorate at KSC is responsible for conducting the testing and evaluation of candidate coatings for inclusion in the APL and for submitting updates to the KSC Engineering Directorate, which is responsible for this list.

Section I. Materials With Greater Than 400 Grams/Liter (3.3 Pounds/Gallon) VOC (SB is Solvent-Based and WB is Water-Based):

Coating Designation	Type	Manufacturer
Dimetcote 9	SB	PPG Industries, Inc. One PPG Place Pittsburgh, PA 15272 (800) 722-4509 http://ppgamercoatus.ppgpmc.com/
Carbo-Zinc 11	SB	Carboline Company 2150 Schuetz Road St. Louis, MO 63146 314-644-1000 www.carboline.com
Cathacoat 304K Cathacoat 304L	SB SB	International Paint LLC/ Devoe Coatings 6001 Antoine Drive Houston, TX 77091(713) 682-1711 (800) 654-2616
Metalhide 1001	SB	PPG Industries, Inc. One PPG Place Pittsburgh, PA 15272 (800) 722-4509 www.ppg.com
ZincClad II	SB	Sherwin-Williams Company 101 Prospect Avenue N.W. Cleveland, OH 44115 (800) 336-1110 www.sherwin-williams.com

Section II. Materials With Less Than 400 Grams/Liter (3.3 Pounds/Gallon) VOC (SB is Solvent-Based and WB is Water-Based):

Coating Designation	Type	Manufacturer
Dimetcote D-9HS Dimetcote D-9H	SB SB	PPG Industries, Inc. One PPG Place Pittsburgh, PA 15272 (800) 722-4509 http://ppgamercoatus.ppgpmc.com/
Carbo-Zinc 11HS Carbo-Zinc 11WB Carbo-Zinc 11 VOC	SB WB SB	Carboline Company 2150 Schuetz Road St. Louis, MO 63146 314-644-1000 www.carboline.com
Cathacoat 305 Cathacoat 304V	WB SB	International Paint LLC/Devoe Coatings Co. 6001 Antoine Drive Houston, TX 77091 (713) 682-1711 (800) 654-2616
InterZinc 22HS	SB	International Paint LLC 6001 Antoine Drive Houston, TX 77091 (713) 682-1711 www.international-pc.com
Zinc Clad XI Zinc Clad II Plus	WB SB	Sherwin-Williams Company 101 Prospect Avenue N.W. Cleveland, OH 44115 (800) 336-1110 www.sherwin-williams.com
Kolor-Zinc 2.8 VOC	SB	Keeler & Long/PPG 856 Echo Lake Road Watertown, CT 06795 1-800-238-8596 http://www.ppg.com/coatings/pmc/brands/keelerlong

APPENDIX B

APPROVED PRODUCTS LIST FOR TOPCOAT SYSTEMS

This list shall be used by or for the Government in the procurement of products covered by this NASA Technical Standard, and such listing of a product is not intended to and does not connote endorsement of the product by NASA. All products listed herein have been tested and meet the requirements for the product as specified. This list is subject to change without notice; revisions or amendments of this list will be issued as necessary. The listing of a product does not release the supplier from compliance with the specification requirements. This list is arranged in two sections based on the coating material's VOC. Use of the information shown herein for advertising or publicity purposes is strictly forbidden.

Thinners and cleaners for each of these coatings shall be procured from the manufacturer of the coating in accordance with sections 4.4.3.1 and 4.4.3.4.

The Materials Test and Corrosion Control Branch in the Engineering Directorate at KSC is responsible for conducting the testing and evaluation of candidate coatings for inclusion in the APL and for submitting updates to the KSC Engineering Directorate, which is responsible for this list.

Section I. Materials With Greater Than 400 Grams/Liter (3.3 Pounds/Gallon) VOC (SB is Solvent-Based and WB is Water-Based):

Primer (Type)	Midcoat (Type)	Topcoat (Type)	Manufacturer
Cathacoat 304L (SB) Cathacoat 304L (SB) Cathacoat 304L (SB) Cathacoat 304K (SB)	Devran 201 (SB)* Devran 230 (SB) Devran 201 (SB)* Devran 201 (SB)*	Devthane 359 (SB) Devthane 369 (SB) Devthane 369 (SB) Devthane 379 UVA (SB)	International Paint LLC/Devoe Coatings 6001 Antoine Drive Houston, TX 77091 (713) 682-1711 (800) 654-2616 www.international-pc.com Houston, TX 77091
MetalHide 1001 (SB)	PittGuard 95-245 (SB)*	PittThane 95-812 (SB)	PPG Industries, Inc. One PPG Place Pittsburgh, PA 15272 (800) 722-4509 www.ppg.com
Kolor-Zinc 2.8 VOC (SB)	N/A	Coraflon ADS (SB)	Keeler & Long/PPG 856 Echo Lake Road Watertown, CT 06795 1-800-238-8596 http://www.ppg.com/coatings/pmc/brands/keelerlong
Carbozinc 11 (SB)	N/A	Carbothane 133 LH (SB)	Carboline Co. 2150 Schuetz Road St. Louis, MO 63146 314-644-1000 www.carboline.com
* Can be used as a direct-to-metal primer for stainless steel, aluminum, and other materials in Zone 5a and 5b environments.			

Section II. Materials With Less Than 400 Grams/Liter (3.3 Pounds/Gallon) VOC (SB is Solvent-Based and WB is Water-Based):

Primer (Type)	Midcoat (Type)	Topcoat (Type)	Manufacturer
D-9HS (SB) D-9HS (SB) D-9H (SB) D-9H (SB) D-9H (SB)	Amerlock 400 (SB)* N/A Amercoat 383 (SB) Amerlock 2/400 (SB)* Amerlock 2/400 (SB)*	Amercoat 450HS (SB) PSX700 (SB) PSX1001 (SB) Amercoat 450H (SB) Amercoat 335 (SB)	PPG Industries, Inc. One PPG Place Pittsburgh, PA 15272 (800) 722-4509 http://ppgamercoatus.ppgpmc.com/
CZ-11HS (SB) CZ-11HS (SB) CZ-11HS (SB) CZ-11WB (WB) CZ-11WB (WB) CZ-11WB (WB)	Carboguard 893(SB)* Carbomastic 15(SB)*Carb 893 (SB)* N/A Carboguard 893 (SB)* Carboacrylic 3358 (WB)	Carbothane 134HS (SB) Carboacrylic 3359(WB) Carboxane 2000 (SB) Carboxane 2000 (SB) Carboacrylic 3359(WB) Carboacrylic 3359(WB)	Carboline Co. 2150 Schuetz Road St. Louis, MO 63146 314-644-1000 www.carboline.com
Cathacoat 304V (SB)	Devran 201 H (SB)*	Devthane 379 (SB)	International Paint LLC/Devoe Coatings 6001 Antoine Drive Houston, TX 77091 (713) 682-1711 (800) 654-2616
Zinc Clad XI (WB) Zinc Clad II Plus (SB) Zinc Clad II Plus (SB)	N/A Macropoxy 646-100 (SB)* Macropoxy 646-100 (SB)*	Polysiloxane XLE Hydrogloss WB (WB) Hi-Solids Poly-CA (SB)	Sherwin-Williams 101 Prospect Ave Cleveland, OH 44115 (800) 336-1110 www.sherwin-williams.com
InterZinc 22HS (SB) InterZinc 22HS (SB)	Interseal 670HS (SB)* Interseal 670HS (SB)*	Interfine 979 (SB) Interfine 878 (SB)	International Paint LLC 6001 Antoine Dr. Houston, TX 77091 (713) 682-1711 www.international-pc.com
* Can be used as a direct-to-metal primer for stainless steel, aluminum, and other materials in Zone 5a and 5b environments.			

Section III. Inorganic Topcoat Systems (SB is Solvent-Based and WB is Water-Based):

Primer (Type)	Midcoat (Type)	Topcoat (Type)	Manufacturer
D-9 HS (SB)	N/A	741 (SB) (IOT)	PPG Industries, Inc. One PPG Place Pittsburgh, PA 15272 (800) 722-4509 http://ppgamercoatus.ppgpmc.com
CZ-11 VOC (SB) CZ-11(SB)	N/A N/A	Carbozinc Finish (SB) (IOT) Carbozinc Finish (SB) (IOT)	Carboline Co. 2150 Schuetz Road St. Louis, MO 63146 314-644-1000 www.carboline.com
Cathacoat 304V (SB)	N/A	Devram 701 (SB) (IOT)	International Paint LLC/ Devoe Coatings 6001 Antoine Drive Houston, TX 77091 (713) 682-1711 (800) 654-2616
InterZinc 22HS (SB)	N/A	Intertherm 181 (SB) (IOT)	International Paint LLC 6001 Antoine Dr Houston, TX 77091 (713) 682-1711 www.international-pc.com
Zinc Clad II (SB)	N/A	L03 (SB) (IOT)	Sherwin-Williams 101 Prospect Ave. Cleveland, OH 44115 (800) 336-1110 www.sherwin-williams.com

APPENDIX C

APPROVED PRODUCTS LIST FOR METALLIZED (TSC) SYSTEMS

This list shall be used by or for the Government in the procurement of products covered by this Standard, and such listing of a product is not intended to and does not connote endorsement of the product by NASA. All products listed herein have been tested and meet the requirements for the product as specified. This list is subject to change without notice; revisions or amendments of this list will be issued as necessary. The listing of a product does not release the supplier from compliance with the specification requirements. Use of the information shown herein for advertising or publicity purposes is strictly forbidden.

Thinners and cleaners for each of these coatings shall be procured from the manufacturer of the coating in accordance with sections 4.4.3.1 and 4.4.3.4.

The Materials Test and Corrosion Control Branch in the Engineering Directorate at KSC is responsible for conducting the testing and evaluation of candidate coatings for inclusion in the APL and for submitting updates to the KSC Engineering Directorate, which is responsible for this list.

Section I. Materials With Greater Than 400 Grams/Liter (3.3 Pounds/Gallon) VOC (SB is Solvent-Based and WB is Water-Based): N/A

Section II: Materials With Less Than 400 Grams/Liter (3.3 Pounds/Gallon) VOC (SB is Solvent-Based and WB is Water-Based):

TSC Primer	Intermediate	Topcoat	Manufacturer
Pure Zinc Pure Zinc Pure Zinc	Interseal 1100 (WB) N/A N/A	Interthane 2100 (SB) Intercryl 520 (WB) Intertherm 181 (SB) (IOT)	International Paint LLC 6001 Antoine Dr Houston, TX 77091 (713) 682-1711 www.international-pc.com
Pure Zinc Pure Zinc Pure Zinc	Macropoxy 646-100 (SB)* Macropoxy 646-100 (SB)*	0 VOC Acrylic (WB) Hi-solids Poly 100 (SB) L03 (SB) (IOT)	Sherwin-Williams 101 Prospect Ave. Cleveland, OH 44115 (800) 336-1110 www.sherwin-williams.com
Pure Zinc Pure Zinc Pure Zinc	N/A N/A N/A	Carbothane 134MC (SB) Carbothane 133MC (SB) Carbozinc Finish (SB) (IOT)	Carboline Co. 2150 Schuetz Road St. Louis, MO 63146 314-644-1000 www.carboline.com

TSC Primer	Intermediate	Topcoat	Manufacturer
Pure Zinc	N/A	Noxyde (WB)	Rustoleum 11 Hawthorn Parkway Vernon Hills, IL 60061 (800) 323-3584 http://www.rustoleumibg.com/
Pure Zinc	N/A	Sky White Powder	Dupont Powder Coatings 9800 Genard Rd Houston, TX 77041 (800) 247-3886 http://www2.dupont.com/Powder/en_US/
Pure Zinc Pure Zinc	BarRust 231V (SB)*	Devthane 379H(SB) Devram 701 (SB) (IOT)	International Paint LLC/Devoe Coatings 6001 Antoine Drive Houston, TX 77091 (713) 682-1711 (800) 654-2616
Pure Zinc Pure Zinc Pure Zinc Pure Zinc	Amerlock Sealer (SB) Amercoat 351 (WB) Amerlock 2 VOC (SB)* N/A	PSX700 (SB) PSX700 (SB) Amershield VOC (WB) 741 (SB) (IOT)	PPG Industries, Inc. One PPG Place Pittsburgh, PA 15272 (800) 722-4509 http://ppgamercoatus.ppgpmc.com
Can be used as a direct-to-metal primer for stainless steel, aluminum, and other materials in Zone 5a and 5b environments.			

APPENDIX D

COATING SPECIFICATION KEY ELEMENTS

1. SCOPE.
2. APPLICABLE DOCUMENTS.
3. SUBMITTALS.
4. ENVIRONMENTAL PROTECTION.
5. WASTE MANAGEMENT.
6. SAFETY/PERSONNEL PROTECTION.
7. MATERIALS.
8. TOOLS AND EQUIPMENT.
9. ENVIRONMENTAL CONDITIONS.
10. WORK SCHEDULE.
11. SURFACE PREPARATION.
12. COATING SCHEDULE (see next page).
13. COATING MIXING AND APPLICATION.
14. QUALITY CONTROL INSPECTION.
15. REPORTING.
16. FINAL ACCEPTANCE.

APPENDIX E

COATING SCHEDULE

SURFACE DESCRIPTION (STEEL, CONCRETE, SURFACE A, B, ETC.)	APPROXIMATE SURFACE AREA SQUARE FEET	SURFACE PREPARATION TYPE	PROFILE RANGE	FIRST COAT			SECOND COAT			THIRD COAT		
				TYPE	WFT/ DFT	COLOR	TYPE	WFT/ DFT	COLOR	TYPE	WFT/ DFT	COLOR

APPENDIX F - COATING SYSTEM DAILY INSPECTION REPORT

COATING SYSTEM DAILY INSPECTION REPORT

DATE	REPORT NO.	PROJECT REF. NO.	PAGE OF

PROJECT DESCRIPTION	LOCATION	CONTRACTOR
INSPECTING ORGANIZATION	INSPECTOR	APPLICABLE SPECIFICATION NO.

I. DESCRIPTION OF ITEMS AND/OR AREAS

II. DESCRIPTION OF WORK PERFORMED/REMARKS

III. PRE-WORK SURFACE CONDITION

- ☐ SUBSTRATE ____
- ☐ GENERAL DESCRIPTION ____
- ☐ PRIMED FOR SUBSEQUENT COATS: REFERENCE REPORT DATED ____
- ☐ PREVIOUSLY PAINTED. DEGREE OF CORROSION ____
- ☐ NEW METAL. DEGREE OF CORROSION ____

OBSERVED DEFECTS

		CORRECTED
OIL & GREASE	☐	☐
SHARP EDGES	☐	☐
WELD SPATTER	☐	☐
MOISTURE	☐	☐
LAMINATIONS	☐	☐
SOLUBLE SALTS	☐	☐
____	☐	☐
____	☐	☐

IV. ENVIRONMENTAL CONDITIONS

TIME	:	:	:	:
AIR TEMP °F				
WET BULB TEMP °F				
RELATIVE HUMIDITY	%	%	%	%
DEW POINT °F				
SURFACE TEMP MIN/MAX °F	/	/	/	/
WIND DIRECTION				
WIND SPEED (MPH)				

REMARKS

V. SURFACE PREPARATION

- ☐ SOLVENT CLEAN ____
- ☐ HAND TOOL ____
- ☐ POWER TOOL ____
- ☐ HP WATER WASH ____
- ☐ ____
- ☐ ____

☐ **ABRASIVE BLAST**

ABRASIVE TYPE ____
BLAST NOZZLE PRESSURE ____
SURFACE PROFILE (AVG) ____
DEGREE OF SURFACE CLEANLINESS ____
COMPRESSED AIR CLEANLINESS ____

START TIME ___:___ STOP TIME ___:___
APPROXIMATE SQ. FT. PREPARED ____
REMARKS ____

PROFILE EFFECT ON TYPE ☐ 1 ☐ 2 GAUGE ____ mils

VI. PRODUCT/MIXING

COATING PRODUCT TYPE ____ MANUFACTURER ____ CATALOG NO./NAME ____ COLOR ____

COATING BATCH NUMBERS
(A) ____
(B) ____
(C) ____

THINNING
THINNER ____
BATCH NO. ____
QTY ADDED ____
% BY VOLUME ____

CAULKING
TYPE ____
MFG ____
PRODUCT NO. ____
BATCH NO. ____

- ☐ **FIRST COAT**
- ☐ **SECOND COAT**
- ☐ **THIRD COAT**
- ☐ ____

TIME MIXED ____ : ____
KIT SIZE ____
GALS MIXED ____
CONTAINER CONDITION ____
PROPERLY STORED? ____
MIXING INSTRUMENT ____
MATERIAL TEMP °F ____

REMARKS ____

VII. COATING APPLICATION

METHOD OF APPLICATION ____
EQUIPMENT DESCRIPTION ____
ATOMIZING AIR CLEANLINESS ____
BRUSHED STRIPE COAT APPLIED TO HARD TO COAT AREAS? ____

START TIME ___:___ STOP TIME ___:___
APPROXIMATE SQ. FT. COATED ____
GALS COATING APPLIED ____
REMARKS ____
WET FILM THICKNESS (AVG) ____ MILS

VIII. POST CURE INSPECTION

☐ DFT WORKSHEET ATTACHED	GAUGE READING	ACTUAL	DATE VERIFIED
SURFACE EFFECT ON GAUGE		N/A	
TOTAL DFT FROM PREVIOUS COATS (AVG)			
DFT THIS COAT (AVG)			

GENERAL APPEARANCE/REMARKS ____

IX. NON-CONFORMANCE ITEMS

DESCRIPTION OF DEFECT	DEFECTIVE ITEMS/AREAS	SPECIFICATION REF. SECTION	N.C.R. NO.	DATE CORRECTED

INSPECTOR'S SIGNATURE DATE

KSC FORM 28-589 (REV. 8/95)

APPENDIX F - COATING SYSTEM DAILY INSPECTION REPORT (Continued)

COATING SYSTEM DAILY INSPECTION REPORT (CONTINUATION SHEET)			
DATE	REPORT NO.	PROJECT REF. NO.	PAGE OF

TIME	REMARKS

SAMPLE

INSPECTOR'S SIGNATURE DATE

KSC FORM 28-589A (3/91)

APPENDIX G - DRY FILM THICKNESS MEASUREMENT WORKSHEET

DRY FILM THICKNESS MEASUREMENT WORKSHEET

DATE	REPORT NO.	PROJECT REF. NO.	APPLICABLE SPECIFICATION	PAGE OF

ITEM/AREA DESCRIPTION	SPOT	SPOT READINGS 1	2	3	TOTAL	AVG (÷3)	% MIN
	A						
	B						
	C						
APPROX SQ. FT.	D						
	E						

SPECIFIED DFT ______ - ______ MILS

RANGE ACHIEVED ______ - ______ MILS

TOTAL

AVG (÷5)

REFERENCE REPORT DATED ______ FOR APPLICATION RECORD

ITEM/AREA DESCRIPTION	SPOT	SPOT READINGS 1	2	3	TOTAL	AVG (÷3)	% MIN
	A						
	B						
	C						
APPROX SQ. FT.	D						
	E						

SPECIFIED DFT ______ - ______ MILS

RANGE ACHIEVED ______ - ______ MILS

TOTAL

AVG (÷5)

REFERENCE REPORT DATED ______ FOR APPLICATION RECORD

ITEM/AREA DESCRIPTION	SPOT	SPOT READINGS 1	2	3	TOTAL	AVG (÷3)	% MIN
	A						
	B						
	C						
APPROX SQ. FT.	D						
	E						

SAMPLE

SPECIFIED DFT ______ - ______ MILS

RANGE ACHIEVED ______ - ______ MILS

TOTAL

AVG (÷5)

REFERENCE REPORT DATED ______ FOR APPLICATION RECORD

ITEM/AREA DESCRIPTION	SPOT	SPOT READINGS 1	2	3	TOTAL	AVG (÷3)	% MIN
	A						
	B						
	C						
APPROX SQ. FT.	D						
	E						

SPECIFIED DFT ______ - ______ MILS

RANGE ACHIEVED ______ - ______ MILS

TOTAL

AVG (÷5)

REFERENCE REPORT DATED ______ FOR APPLICATION RECORD

ITEM/AREA DESCRIPTION	SPOT	SPOT READINGS 1	2	3	TOTAL	AVG (÷3)	% MIN
	A						
	B						
	C						
APPROX SQ. FT.	D						
	E						

SPECIFIED DFT ______ - ______ MILS

RANGE ACHIEVED ______ - ______ MILS

TOTAL

AVG (÷5)

REFERENCE REPORT DATED ______ FOR APPLICATION RECORD

ITEM/AREA DESCRIPTION	SPOT	SPOT READINGS 1	2	3	TOTAL	AVG (÷3)	% MIN
	A						
	B						
	C						
APPROX SQ. FT.	D						
	E						

SPECIFIED DFT ______ - ______ MILS

RANGE ACHIEVED ______ - ______ MILS

TOTAL

AVG (÷5)

REFERENCE REPORT DATED ______ FOR APPLICATION RECORD

REMARKS

TOTAL SQUARE FOOTAGE COATED (APPROX)	INSPECTOR'S SIGNATURE	DATE

KSC FORM 28-588 (3/91)

APPENDIX H

DELIVERABLES

H.1 GSE Documentation Deliverables

a. The GSE provider shall submit documentation to verify that the hardware/software has been developed in accordance with this NASA Technical Standard.

b. The GSE provider shall provide all documentation to the using organization when the GSE is delivered for use, regardless of who "owns" the GSE at the time of delivery.

Examples of this documentation include, but are not limited to, the following:

(1) *Certification Approval Request (indicates how the GSE was certified as complying with this NASA Technical Standard).*

(2) *Master Verification Matrix (indicates which GSE requirements were met and how).*

(3) *Material Inspection and Receiving Report.*

(4) *Validation and verification compliance records.*

(5) *Drawings with parts list or bills of material.*

(6) *Maintenance manuals/procedures.*

(7) *Material certifications and lot traceability.*

(8) *Operating manuals/procedures.*

(9) *Software Version Description document.*

(10) *Firmware Version Description document.*

(11) *Facility and Flight Vehicle Interface requirements.*

(12) *Hazard Analyses or Ground Safety Data pack.*

(13) *Failure Modes, Effects, and Criticality Analysis.*

(14) *Critical Items List.*

c. *Intent/Rationale: The using organization requires documentation for safely operating, maintaining, and servicing the GSE.*

d. To reduce risk to the mission as well as to ground personnel and flight crews, the GSE provider shall complete and submit a failure mode and effects analysis in accordance with the criticality assigned to the GSE by the responsible program or project.

APPENDIX I

GUIDANCE

I.1 Purpose and/or Scope

The purpose of this Appendix is to provide guidance and is made available in the reference documents listed below.

I.2 Reference Documents

ASTM

ASTM D4228	Standard Practice for Qualification of Coating Applicators for Application of Coatings to Steel Surfaces

Code of Federal Regulations (CFR)

29 CFR Part 1910	Occupational Safety and Health Administration (Occupational Safety and Health Standards)
29 CFR Part 1926	Occupational Safety and Health Administration (Safety and Health Regulations for Construction)

Compressed Gas Association, Inc. (CGA)

CGA G-7.1	Commodity Specification for Air

Department of Defense

MIL-P-85891	Plastic Media for Removal of Organic Coatings
TO 1-1-691	Cleaning and Corrosion Prevention and Control, Aerospace and Non-Aerospace Equipment

NASA

KSC-TM-584	Corrosion Control and Treatment Manual

The Society for Protective Coatings (SSPC)

SSPC-SP 6/NACE No. 3	Commercial Blast Cleaning
SSPC-SP 7/NACE No. 4	Brush-off Blast Cleaning
SSPC-VIS 1-89	Visual Standard for Abrasive Blast-Cleaned Steel
SSPC-VIS 3	Visual Reference for Power and Hand Tool Cleaned Steel

I.3 Additional Related Information

For information and guidance on dissimilar metals, corrosion-inhibiting lubricants, etc., refer to TO 1-1-691, Cleaning and Corrosion Prevention and Control, Aerospace and Non-Aerospace Equipment; and KSC-TM-584, Corrosion Control and Treatment Manual.

APPENDIX J

REQUIREMENTS COMPLIANCE MATRIX

A.1 Purpose

This Appendix provides a listing of requirements contained in this NASA Technical Standard for selection and verification of requirements by programs and projects. *(**Note**: Enter "Yes" to describe the requirement's applicability to the program or project; or enter "No" if the intent is to tailor, and enter how tailoring is to be applied in the "Rationale" column.)*

NASA-STD-NASA-STD-5008B w/CHANGE 1				
Section	**Description**	**Requirement in this NASA Technical Standard**	**Applicable (Yes or No)**	**If No, Enter Rationale**
1.2.4	Applicability	[PCR 1] This NASA Technical Standard shall be used in the preparation of written, individual coating specifications for specific projects for the prevention of corrosion through the use of protective coatings on space vehicle launch structures, facilities, GSE, and test facilities and structures in the specific environments identified in section 1.4.		
1.2.5	Applicability	[PCR 2] Due to the changing environmental considerations and different site conditions, new advances in corrosion technology, and a wide array of possible applications, this NASA Technical Standard shall not be used as a stand-alone specification that meets every contingency.		
1.2.6	Applicability	[PCR 3] The appendices are considered to be an integral part of this NASA Technical Standard. Appendices A, B, C, and D shall be used for the preparation of all coating specifications.		
1.3	Tailoring	[PCR 4] Tailoring of this NASA Technical Standard for application to a specific program or project shall be formally documented as part of program or project requirements and approved by the Technical Authority in accordance with NPR 7120.5, NASA Space Flight Program and Project Management Requirements.		
1.4	Zones of Exposure	[PCR 5] The zones of exposure established to define coating system requirements for surfaces located in specific environments shall be determined by the Design		

		Engineer responsible for preparing the coating specification from the following zones: a. (1) Zone 1a. Surfaces that are directly impinged on by solid rocket booster (SRB) engine exhaust. (2) Zone 1b. Surfaces that are indirectly impinged on by SRB exhaust. (3) Zone 1c. Walking surfaces in Zones 1a and 1b. b. Zone 2. Surfaces that are exposed to elevated temperatures (above 65 °C (above 150 °F)) and/or acid deposition from SRB exhaust with no exhaust impingement. c. (1) Zone 3a. Surfaces, other than those located in Zones 1 or 2, that are exposed to acid deposition from SRB exhaust products. (2) Zone 3b. Surfaces that are exposed to other types of chemical contamination (e.g., cooling towers, diesel exhaust stacks, acidic industrial environments, and water treatment facilities). d. (1) Zone 4a. Surfaces not located in the launch environment but located in a neutral pH corrosive marine industrial environment or other chloride-containing environments. (2) Zone 4b. Surfaces located in neutral pH exterior environments in any geographical area. (3) Zone 4c. Surfaces located in indoor environments that are not air-conditioned. e. (1) Zone 5a. Surfaces located in a continuous indoor air-conditioned environment, such as an office or clean room, where both temperature and humidity are controlled more than 90 percent of the time. (2) Zone 5b. Surfaces located in a low humidity, high ultraviolet environment, such as a high altitude, arid location. f. (1) Zone 6a. Surfaces located underground or subject to intermittent or continuous immersion in aqueous environments. (2) Zone 6b. Surfaces subject to exposure in a chemical/fuel storage environment. Based on the complexity of the liquid stored, this has to be engineered separately in compliance with all federal, state, and local environmental statutory requirements. g. Zone 7. Surfaces under thermal insulation, such as chilled water, steam, and heated gas lines.		
1.5a	Method of Specifying Coating Requirements	[PCR 6] Specifications referencing this NASA Technical Standard shall include the following:		

		(1) The type of surface to be coated. (2) The zone of exposure. (3) Surface preparation. (4) Defined paint system. (5) Coating thicknesses. (6) The finish color required (when applicable).		
1.5b	Method of Specifying Coating Requirements	[PCR 7] The coating specification shall contain the following key elements: scope, applicable documents, submittals, environmental protection, waste management, safety/personnel protection, materials, tools and equipment, environmental conditions, work schedule, surface preparation (including a listing of abrasive-sensitive hardware to be prepared or protected), coating schedule, coating mixing and application, quality control inspection, reporting, and final acceptance.		
1.6a	Environmental Stewardship and Health and Safety	[PCR 8] Environmental, health, and safety impacts of processes and materials shall be taken into account when employing protective coating methods and techniques.		
1.6b	Environmental Stewardship and Health and Safety	[PCR 9] Alternative, environmentally friendly materials that do not contain hexavalent chromium, lead, cadmium, or hazardous air pollutants (HAPs), such as methyl ethyl ketone, toluene, and xylene, shall be considered when determining the correct coating method/technique for each protective coating application.		
1.6c	Environmental Stewardship and Health and Safety	[PCR 10] Alternative, less hazardous materials shall be considered when determining the correct coating method/technique for each protective coating application to minimize risk in construction, use, and demolition.		
1.6d	Environmental Stewardship and Health and Safety	[PCR 11] Performance criteria defined in this NASA Technical Standard shall take precedent.		
2.1.1	General	[PCR 12] The latest issuances of cited documents shall apply unless specific versions are designated		
2.1.2	General	[PCR 13] Non-use of specifically designated versions shall be approved by the responsible Technical Authority.		
2.4.2	Order of Precedence	[PCR 14] Conflicts between this NASA Technical Standard and other requirements documents shall be resolved by the responsible Technical Authority.		
4.1.1a	Abrasive-Blasting Aggregate	[PCR 15] Blasting aggregates shall be approved materials in accordance with MIL-A-22262, Abrasive Blasting Media Ship Hull Blast Cleaning; or SSPC-AB 1, Mineral and Slag Abrasives, Type I or II, Class A; or steel grit in accordance with SSPC-AB 3, Ferrous Metallic Abrasive, Class 1.		
4.1.1b	Abrasive-Blasting Aggregate	[PCR 16] Only materials approved in Qualified Products List (QPL) 22262, Qualified Products List: List of Products Qualified Under Military Specification MIL-A-22262, shall be used.		

4.1.1c	Abrasive-Blasting Aggregate	[PCR 17] The abrasive grade selected shall produce the required surface profile and possess physical properties that are compatible with the requirements of this NASA Technical Standard.		
4.1.1d	Abrasive-Blasting Aggregate	[PCR 18] The new steel grit shall be a neutral pH (6.0 to 8.0), rust-free and oil-free, dry, commercial-grade blasting grit with a hardness of 40 to 50 Rockwell C.		
4.1.1e	Abrasive-Blasting Aggregate	[PCR 19] Recycled steel grit shall be in accordance with SSPC-AB 2, Cleanliness of Recycled Ferrous Metallic Abrasives.		
4.1.1f	Abrasive-Blasting Aggregate	[PCR 20] The size shall be selected to produce the required anchor profile.		
4.1.1g	Abrasive-Blasting Aggregate	[PCR 21] Only aggregates that are free of crystalline silica shall be selected for use at NASA unless exemptions to this policy are coordinated with the local Occupational Health Office.		
4.1.1h	Abrasive-Blasting Aggregate	[PCR 22] Blasting aggregate for abrasion-sensitive hardware (such as bellows, gimbal joints, and other thin-walled components) shall be materials that do not change the surface profile.		
4.1.1h(1)	Abrasive-Blasting Aggregate	[PCR 23] Blasting operations shall not produce holes, cause distortion, remove metal, or cause thinning of the substrate.		
4.1.2a	Protective Coatings, Thinners, and Cleaners	[PCR 24] All coatings shall possess physical properties and handling characteristics that are compatible with the application requirements of this NASA Technical Standard.		
4.1.2b	Protective Coatings, Thinners, and Cleaners	[PCR 25] All coatings shall be self-curing.		
4.1.2c	Protective Coatings, Thinners, and Cleaners	[PCR 26] Thinners and cleaners for each coating shall be procured from the manufacturer of the coating.		
4.1.2d	Protective Coatings, Thinners, and Cleaners	[PCR 27] Procurement awards for coatings to be supplied according to this NASA Technical Standard shall be made only for those products in the Approved Products List (APL).		
4.1.2e	Protective Coatings, Thinners, and Cleaners	[PCR 28] Application characteristics shall be judged acceptable prior to beach exposure testing.		
4.1.2f	Protective Coatings, Thinners, and Cleaners	[PCR 29] Protective coatings shall be compatible with fluids expected in the areas to the extent required to prevent fire, explosion, or damage to facility, hardware, and GSE.		
4.1.2g	Protective Coatings, Thinners, and Cleaners	[PCR 30] All coating materials, when used in areas where exposure to hypergolic propellants could occur, shall be compatible with the propellants in accordance with		

		NASA-STD-6001, Flammability, Offgassing, and Compatibility Requirements and Test Procedures.		
4.1.2.1a	Inorganic Zinc Coatings	[PCR 31] To be listed, a coating shall meet the following minimum requirements: (1) Self-curing, multiple-component. (2) Dry-temperature resistance to 400 °C (750 °F) for 24 hours. (3) Minimum shelf life of 6 months when stored in accordance with manufacturer's instructions. (4) Minimum of 83 percent zinc by weight in the applied dry film. (5) Contain Type III zinc dust pigment in accordance with ASTM D520, Standard Specification for Zinc Dust Pigment, and be asbestos-free, polychlorinated biphenyl (PCB)-free, lead-free, cadmium-free, and chromate-free (less than 0.002 percent by weight of mixed coating). (6) Attain a rating of not less than 9 in accordance with ASTM D610, Standard Practice for Evaluating Degree of Rusting on Painted Steel Surfaces; and ASTM D1654, Standard Test Method for Evaluation of Painted or Coated Specimens Subjected to Corrosive Environments, when applied to composite carbon steel test panels and exposed at the KSC Beach Corrosion Test Site for the following periods: A. 18 months for initial acceptance. B. 5 years for final acceptance.		
4.1.2.1b	Inorganic Zinc Coatings	[PCR 32] Application characteristics shall be judged acceptable prior to beach testing.		
4.1.2.2.1	Inhibitive Polyamide Epoxy Coatings	[PCR 33] Polyamide epoxy coatings shall conform to the following minimum requirements: a. Polyamide-cured. b. Rust-inhibitive. c. PCB-free, lead-free, cadmium-free, and chromate-free (less than 0.002 percent by weight of mixed coating). d. Suitable as a primer for carbon steel, galvanized steel, and aluminum. e. Suitable as an intermediate coat between an inorganic zinc primer and an aliphatic polyurethane finish coat. f. Meet the compatibility requirements of section 4.4.3.1. g. Contain a minimum of 40 percent solids by volume.		

4.1.2.2.2	Noninhibitive Polymide Epoxy Coatings	[PCR 34] Polyamide epoxy coatings shall conform to the following minimum requirements: a. Polyamide-cured. b. PCB-free, lead-free, cadmium-free, and chromate-free (less than 0.002 percent by weight of mixed coating). c. Suitable as an intermediate coat between inorganic zinc primer and an aliphatic polyurethane finish coat. d. Meet the compatibility requirements of section 4.4.3.1. e. Contain a minimum of 40 percent solids by volume. f. Not to be used as a primer on carbon steel.		
4.1.2.2.3	Water-Reducible Intermediate Coatings	[PCR 35] Water-reducible intermediate coatings shall conform to the following minimum requirements: a. Self-curing, one or two packages, water reducible. b. PCB-free, lead-free, cadmium-free, and chromate-free (less than 0.002 percent by weight of mixed coating). c. Suitable as an intermediate coat between inorganic zinc primers and water-reducible topcoats. d. Meet the compatibility requirements of section 4.4.3.1. e. Contain a minimum of 30 percent solids by volume. f. Not to be used as a primer on steel.		
4.1.2.3.1	Aliphatic Polyurethane Coatings	[PCR 36] Aliphatic polyurethane coatings shall conform to the following minimum requirements: a. Catalyst isocyanate cured. b. High-gloss finish (minimum 85 gloss units (GUs) at a 60-degree angle). c. Retain gloss and color upon prolonged exterior exposure. d. Suitable as an exterior finish coat over an inorganic zinc primer with a polyamide epoxy intermediate coat. e. Meet the compatibility requirements of section 4.4.3.1. f. Contain a minimum of 44 percent solids by volume. g. PCB-free, lead-free, cadmium-free, and chromate-free (less than 0.002 percent by weight of mixed coating). h. Attain a numerical rating of not less than 8 in accordance with ASTM D610 and ASTM D1654 and a numerical rating of not less than 9F in accordance with ASTM D714, Standard Test Method for Evaluating Degree of Blistering of Paints, when applied as a system to composite carbon steel test panels and exposed at the KSC Beach Corrosion Test Site for the following periods:		

		(1) 18 months for initial acceptance. (2) 5 years for final acceptance.		
4.1.2.3.2	Water-Reducible Topcoats	[PCR 37] Water-reducible topcoats shall conform to the following minimum requirements: a. Self-curing, one or two packages, water-reducible. b. PCB-free, lead-free, cadmium-free, and chromate-free (less than 0.002 percent by weight of mixed coating). c. Retain gloss and color upon prolonged exterior exposure. (1) Semi-gloss or high-gloss finish. (Semi-gloss is defined as 60 GU to 85 GU at a 60-degree angle; high gloss is defined as a minimum 85 GU at a 60-degree angle.) e. Meet the compatibility requirements of section 4.4.3.1. f. Attain a numerical rating of not less than 8 in accordance with ASTM D610 and ASTM D1654 and a numerical rating of not less than 9F in accordance with ASTM D714, when applied as a system to composite carbon steel test panels and exposed at the KSC Beach Corrosion Test Site for the following periods: (1) 18 months for initial acceptance. (2) 5 years for final acceptance.		
4.1.2.3.3	Inorganic Topcoats (IOTs)	[PCR 38] IOTs shall conform to the following minimum requirements: a. Dry-temperature resistance to 400 °C (750 °F) for 24 hours. b. Suitable as a topcoat for inorganic zinc and galvanized steel in high-temperature environments. c. Listed as an approved coating system (see Appendix B). d. PCB-free, lead-free, cadmium-free, and chromate-free (less than 0.002 percent by weight of mixed coating). e. Attain a rating of not less than 9 in accordance with ASTM D610 and ASTM D1654 when applied to composite carbon steel test panels and exposed at the KSC Beach Corrosion Test Site for the following periods: (1) 18 months for initial acceptance. (2) 5 years for final acceptance.		
4.1.2.3.4	Polysiloxane Topcoats	[PCR 39] Polysiloxane topcoats shall conform to the following minimum requirements: a. Suitable as a finish coat for exterior exposure. b. Contain a minimum of 44 percent solids by volume. c. High-gloss finish (minimum 85 GU at a 60-degree angle). d. Retain gloss and color on prolonged outdoor exposure.		

		e. PCB-free, lead-free, cadmium-free, and chromate-free (less than 0.002 percent by weight of mixed coating). f. Listed as an approved coating system (see Appendix B). g. Attain a numerical rating of not less than 8 in accordance with ASTM D610 and ASTM D1654 and a numerical rating of not less than 9F in accordance with ASTM D714 when applied as a system to composite carbon steel test panels and exposed at the KSC Beach Corrosion Test Site for the following periods: (1) 18 months for initial acceptance. (2) 5 years for final acceptance.		
4.1.2.4	Epoxy Mastic Coatings	[PCR 40] Epoxy mastic coatings shall conform to the following minimum requirements: a. Specifically intended for use over mechanically cleaned steel. b. Contain a minimum of 80 percent solids by volume. c. Two-component, catalyst-cured, aluminum-pigmented. d. PCB-free, lead-free, cadmium-free, and chromate-free (less than 0.002 percent by weight of mixed coating).		
4.1.2.5a	Coal Tar Epoxy	[PCR 41] Coal tar epoxy coating shall be a two-component, high-build epoxy of low volatile organic content (VOC).		
4.1.2.5b	Coal Tar Epoxy	[PCR 42] The coal tar epoxy shall contain, at a minimum, 65 percent solids by volume.		
4.1.2.5c	Coal Tar Epoxy	[PCR 43] The coal tar epoxy shall produce a one-coat thickness of 405 µm to 510 µm (16 mil to 20 mil) per coat dry film thickness (DFT).		
4.1.2.6	Potable Water Epoxy	[PCR 44] All coatings for potable water immersion service shall be three-coat epoxy systems that are certified by NSF Standard 61, Drinking Water System Components – Health Effects.		
4.1.2.7	Nonskid Coating	[PCR 45] Approved nonskid coatings shall conform to MIL-PRF-24667, Coating System, Non-skid, for Roll, Spray, or Self-Adhering Application, Type 1, Composition G, as supplied by American Safety Technologies, Inc., 565 Eagle Rock Avenue, Roseland, NJ 07068, telephone (800) 631-7841 (< www.astantislip.com >), or an approved equivalent (Primer MS 7C, Topcoat MS 400G, Color Topping MS-200).		
4.1.3a	Sealants/Caulking	[PCR 46] Sealants shall be self-curing, single-component, polysulfide rubber or polyurethane material only, conforming to ASTM C920, Standard Specification for Elastomeric Joint Sealants, Type S, Grade NS, Class 25, use NT, A, and O.		
4.1.3b	Sealants/Caulking	[PCR 47] If not topcoated, the caulking shall match the color of the joint surface being caulked.		

4.1.3c	Sealants/Caulking	[PCR 48] If caulking is to be used in a clean-room environment, an approved material with low offgassing characteristics in accordance with NASA-STD-6001 shall be selected.		
4.1.4	Chip-Free Clean-Room Paint	[PCR 49] Offgassing, flammability, and hypergolic compatibility testing shall be in accordance with NASA-STD-6001, Supplemental Test Procedure A.7.		
4.2.1a	Compressed Air	[PCR 50] The compressed air system shall be capable of delivering a continuous nozzle pressure to achieve the required surface cleanliness and profile, typically 620 kPa (90 psi) minimum to each blast nozzle in operation.		
4.2.1b	Compressed Air	[PCR 51] The compressed air system shall be equipped with oil and moisture separators to ensure only clean, dry air is provided to the service outlet.		
4.2.1c	Compressed Air	[PCR 52] The compressed air system shall comply with Occupational Safety and Health Administration (OSHA), American National Standards Institute (ANSI), and National Institute of Occupational Safety and Health (NIOSH) configurations.		
4.2.1d	Compressed Air	[PCR 53] Air distribution manifolds shall conform to American Society of Mechanical Engineers (ASME) standards.		
4.2.2a	Abrasive-Blasting System	[PCR 54] The abrasive-blasting system shall comply with OSHA, ANSI, and NIOSH configurations consisting of, but not limited to, the following: (1) A remote-controlled welded pressure pot conforming to ASME standards. (2) The required length of blast hose. (3) A venturi nozzle. (4) A respiratory air-line filter. (5) A blast hood approved by the Mine Safety and Health Administration/NIOSH with the required length of air hose.		
4.2.2b	Abrasive-Blasting System	[PCR 55] The blasting system shall be designed to produce the specified cleanliness level and profile when coupled with the available compressed air supply.		
4.2.3	Coating Application System	[PCR 56] The coating application equipment shall be an airless spray system, conventional spray system, or other approved equipment in accordance with the coating manufacturer's recommendations and section 4.4.3.6.		
4.2.4	Respiratory Protection	[PCR 57] Respiratory protection shall be in accordance with 29 CFR 1910.134, Respiratory Protection, and Center respiratory protection requirements.		
4.3a	Safety and Health Requirements	[PCR 58] Necessary precautions, in accordance with OSHA regulations, manufacturers' recommendations, and industry standards, shall be taken to ensure the safety and health of personnel performing the work required by this NASA Technical Standard and personnel who may be affected by such work.		
4.3b	Safety and Health Requirements	[PCR 59] The Contractor shall provide equipment as required for safe and healthful application and instruction to the users regarding the hazards and proper handling and disposal procedures to prevent injury or illness.		

4.3c	Safety and Health Requirements	[PCR 60] The Contractor shall provide safe access to all areas for the coating inspector.		
4.3d	Safety and Health Requirements	[PCR 61] The Contractor shall submit a written safety and health plan that includes a Hazard Communication Program, a Respiratory Protection Program, and a Hearing Conservation Program that conforms to OSHA requirements and industry standards.		
4.3e	Safety and Health Requirements	[PCR 62] Where the contractor is required to remove surface coatings that contain PCB, lead, chromium, mercury, or cadmium, or other regulated materials, the Contractor shall include specific provisions in the safety and health plan for complying with all Federal, State, Local, and NASA Center-specific requirements.		
4.3.1a	Environmental Requirements	[PCR 63] All local, state, and federal environmental regulations, as well as the NASA Center's environmental policies, shall be followed.		
4.3.1b	Environmental Requirements	[PCR 64] Questions regarding these regulations and policies shall be directed to the local environmental management organization.		
4.3.2a	Personal Protective Equipment (PPE)	[PCR 65] When engineering controls cannot be implemented to protect workers, then PPE and/or administrative controls shall be used.		
4.3.2b	Personal Protective Equipment (PPE)	[PCR 66] Where required, PPE shall be used in accordance with all federal, state, NASA, and Center requirements.		
4.3.2c	Personal Protective Equipment (PPE)	[PCR 67] Both the supervisors and the workers shall be properly instructed, trained, and certified in the selection, use, and maintenance of PPE.		
4.4.1a	Application Qualifications	[PCR 68] To ensure the highest quality of workmanship, only coating applicators who have worked in the painting trade sufficiently long enough to master the use of all applicable tools and materials shall be assigned to perform the work described herein.		
4.4.1b	Application Qualifications	[PCR 69] In addition, the coating applicators shall provide written evidence of having successfully completed a comprehensive training program, such as Painting and Decorating Contractors of America (PDCA)/NACE/SSPC Industrial Painters Training, or equivalent.		
4.4.1c	Application Qualifications	[PCR 70] The Contractor shall provide all painting personnel an orientation on the proper mixing and application of the coatings specified, particularly inorganic zinc coatings.		
4.4.1d	Application Qualifications	[PCR 71] Topics in the orientation of proper mixing and application of the specified coatings (particularly for inorganic zinc coatings) shall include specification requirements, material application characteristics, and inspection criteria.		
4.4.1e	Application Qualifications	[PCR 72] The mixing or application of coatings shall be performed only by personnel who have received training.		

4.4.1f	Application Qualifications	[PCR 73] The Contractor shall prepare representative sample areas that meet specification requirements.		
4.4.2a	Preparation of Surfaces	[PCR 74] All surfaces to be coated shall be clean, dry, and free from oil, grease, dirt, dust, corrosion, peeling paint, caulking, weld spatter, and any other surface contaminants.		
4.4.2b	Preparation of Surfaces	[PCR 75] All surfaces that cannot be accessed after fabrication, erection, or installation shall be prepared and coated while accessible.		
4.4.2c	Preparation of Surfaces	[PCR 76] Surface preparation and coating operations shall be sequenced, so that freshly applied coatings will not be contaminated by dust or foreign matter.		
4.4.2d	Preparation of Surfaces	[PCR 77] All equipment and adjacent surfaces not to be coated shall be protected from surface preparation operations.		
4.4.2e	Preparation of Surfaces	[PCR 78] Working mechanisms shall be protected against intrusion of the abrasive.		
4.4.2f	Preparation of Surfaces	[PCR 79] All surfaces shall be degreased, as required, before subsequent surface preparation procedures or the application of protective coatings, or both.		
4.4.2g	Preparation of Surfaces	[PCR 80] The following sections provide the surface preparation techniques that shall be used when specified in section 4.5.		
4.4.2.1a	Cleaning and Degreasing	[PCR 81] Degreasing shall be by solvent cleaning, detergent washing, or steam cleaning in accordance with SSPC-SP 1, Solvent Cleaning.		
4.4.2.1b	Cleaning and Degreasing	[PCR 82] This degreasing procedure shall be followed when cleaning carbon steel, galvanized steel, stainless steel, or aluminum.		
4.4.2.1c	Cleaning and Degreasing	[PCR 83] Selection of solvents shall be in accordance with use requirements and applicable federal, state, and NASA environmental policies.		
4.4.2.1d	Cleaning and Degreasing	[PCR 84] Chlorofluorocarbon solvents shall not be used.		
4.4.2.1e	Cleaning and Degreasing	[PCR 85] Water washing, using clean potable water, shall be done when high levels of chloride ($>5\ \mu g/cm^2$) or other undesirable contaminants are found on the surfaces.		
4.4.2.1f	Cleaning and Degreasing	[PCR 86] Water washing shall be accomplished using standard industrial pressure cleaners with a pressure-versus-volume output balance that will ensure thorough and productive cleaning.		
4.4.2.1g	Cleaning and Degreasing	[PCR 87] All water washing or pressure cleaning operations shall comply with all Federal, State, Local, and NASA Center environmental requirements.		
4.4.2.1h	Cleaning and Degreasing	[PCR 88] The cleaned surface shall be free of loose coatings, chlorides, dirt, dust, mildew, grinding/welding/cutting debris, and visible contaminants.		
4.4.2.1i	Cleaning and Degreasing	[PCR 89] The surface shall be clean and dry prior to the abrasive-blasting operations and application of coatings.		

4.4.2.2a	Abrasive Blasting	[PCR 90] The abrasive-blasting aggregate shall be clean and dry and conform to section 4.1.1.		
4.4.2.2b	Abrasive Blasting	[PCR 91] The abrasive-blasting system shall conform to section 4.2.2.		
4.4.2.2c	Abrasive Blasting	[PCR 92] Abrasive blasting shall be in accordance with the applicable paragraphs in section 4.5.		
4.4.2.2d	Abrasive Blasting	[PCR 93] Abrasive residues shall be removed from the surface, leaving it clean and dry before the coatings are applied.		
4.4.2.2e	Abrasive Blasting	[PCR 94] All particulate emissions generated during abrasive-blasting operations shall be contained.		
4.4.2.2f	Abrasive Blasting	[PCR 95] The containment system shall be designed to comply with all applicable federal, state, and local regulations as well as all NASA policies.		
4.4.2.2g	Abrasive Blasting	[PCR 96] Exemptions to the requirement in 4.4.2.2f shall be coordinated with the local environmental management office.		
4.4.2.2h	Abrasive Blasting	[PCR 97] The aggregate used to prepare abrasive-sensitive hardware such as bellows, gimbal joints, and other thin-walled components shall be carefully identified and selected.		
4.4.2.3	Mechanical Cleaning Methods	[PCR 98] Mechanical methods shall be in accordance with the applicable paragraph in section 4.5.		
4.4.3a	Application of Coatings	[PCR 99] All prepared surfaces shall be coated within 6 hours after surface preparation and before corrosion or recontamination occurs. *As an exception, surfaces prepared under temperature and humidity control may be coated after 6 hours but only after inspection of the surface preparation confirms that the cleanliness level has met the specified standards.*		
4.4.3b	Application of Coatings	[PCR 100] Any surface that shows corrosion or contamination, regardless of the length of time after preparation, shall be prepared again.		
4.4.3c	Application of Coatings	[PCR 101] Because the application and handling characteristics of all coatings will vary, adequate written instructions from the manufacturer are essential and shall be closely followed in conjunction with the requirements defined herein to obtain optimum performance.		
4.4.3d	Application of Coatings	[PCR 102] The manufacturer's written recommendations for thinning, mixing, handling, and applying the product shall be strictly followed.		
4.4.3e	Application of Coatings	[PCR 103] All coatings shall be thoroughly worked into all joints, crevices, and open spaces.		
4.4.3f	Application of Coatings	[PCR 104] All newly coated surfaces shall be protected from damage.		
4.4.3g	Application of Coatings	[PCR 105] All equipment and adjacent surfaces not to be coated shall be protected from overspray and splattered coatings.		

4.4.3h	Application of Coatings	[PCR106] Particulate emissions shall be contained during all spray-painting operations.		
4.4.3i	Application of Coatings	[PCR 107] The containment system shall be designed to comply with all federal, state, and local regulations as well as all NASA policies.		
4.4.3j	Application of Coatings	[PCR 108] Exemptions to this requirement shall be coordinated with the local environmental management organization.		
4.4.3.1a	Coating Systems	[PCR 109] Coating systems for specified uses and substrates shall be as defined in section 4.5 and conform to section 4.1.2.		
4.4.3.1b	Coating Systems	[PCR 110] All thinners and cleaners shall be products of the coating manufacturer, except as defined in section 4.1.2.7.		
4.4.3.1c	Coating Systems	[PCR 111] To ensure intercoat compatibility, coating systems consisting of more than one coat shall be products of the same manufacturer.		
4.4.3.1d	Coating Systems	[PCR 112] Continuity of the coating manufacturer's system shall be maintained for the duration of an individual project.		
4.4.3.2a	Colors	[PCR 113] Inorganic zinc coatings shall be pigmented so that there is a definite contrast between the coating and the dull gray appearance of the blasted steel surface during the coating application.		
4.4.3.2b	Colors	[PCR 114] Color coding for fluid system piping shall be in accordance with KSC-STD-SF-0004, Ground Piping Systems Color Coding and Identification, Safety Standard for.		
4.4.3.2c	Colors	[PCR 115] Finish coat colors shall be in accordance with the following FED-STD-595, Colors Used in Government Procurement, color numbers using pigments free of PCB, lead, chromium, and cadmium: (1) White, No. 17925. (2) Blue, No. 15102 (safety). (3) Yellow, No. 13538 (standard). (4) Yellow, No. 13655 (safety). (5) Red, No. 11136. (6) Red, No. 11105 (safety). (7) Black, No. 17038. (8) Green, No. 14110 (safety). (9) Gray, No. 16187 (safety). (10) Brown, No. 10080 (safety). (11) Gray, No. 16473 (standard).		
4.4.3.3a	Storage of Coating Materials	[PCR 116] Coating materials and thinners shall be stored in their original containers with the manufacturer's name, product identification, shelf life, and batch number.		

4.4.3.3b	Storage of Coating Materials	[PCR 117] Coating materials, thinners, and cleaners shall be stored in tightly closed containers in a covered, well-ventilated area where they will not be exposed to sparks, flame, direct sunlight, extreme heat, or rainfall.		
4.4.3.3c	Storage of Coating Materials	[PCR 118] The manufacturer's written instructions for storage limitations shall be followed.		
4.4.3.3d	Storage of Coating Materials	[PCR 119] Tarpaulins shall not be used as the sole means of covering coating materials for storage.		
4.4.3.3e	Storage of Coating Materials	[PCR 120] Material Safety Data Sheets for coating materials and thinners shall be maintained or made accessible to users in the area.		
4.4.3.3f	Storage of Coating Materials	[PCR 121] The Contractor shall submit a written plan for approval for storage of coating materials for coordination with the local safety/fire/environmental organization.		
4.4.3.4a	Mixing and Application Instructions	PCR 122] Coating materials shall be thoroughly mixed prior to application with a mechanical mixing instrument that will not induce air into the coating, such as a Jiffy Mixer, manufactured by the Jiffy Mixer Company (< www.jiffymixer.com >), Inc., Riverside, CA, or an approved equivalent.		
4.4.3.4b	Mixing and Application Instructions	[PCR 123] The mixer shall be powered by an air motor or an explosion-proof electric motor.		
4.4.3.4c	Mixing and Application Instructions	[PCR 124] All mixing operations shall be performed over an impervious surface with provisions to prevent runoff to grade of any spilled material.		
4.4.3.4d	Mixing and Application Instructions	[PCR 125] The mixed coating material shall be strained through a 30-mesh to 60-mesh screen prior to application.		
4.4.3.4e	Mixing and Application Instructions	[PCR 126] Thinning shall be for viscosity control only.		
4.4.3.4f	Mixing and Application Instructions	[PCR 127] If thinner is required, the amount recommended by the manufacturer of the thinner shall be used.		
4.4.3.4g	Mixing and Application Instructions	[PCR 128] The material shall be agitated as required during application to maintain a uniform suspension of solids.		
4.4.3.4h	Mixing and Application Instructions	[PCR 129] Continuous rapid agitation shall be avoided.		

4.4.3.4i	Mixing and Application Instructions	[PCR 130] Spray equipment shall be adjusted to produce an even, wet coat with minimum overspray.		
4.4.3.4j	Mixing and Application Instructions	[PCR 131] The conventional pressure pot, when used, shall be kept at approximately the same level or above the spray gun, so that the material is delivered properly.		
4.4.3.4k	Mixing and Application Instructions	[PCR 132] Coatings shall be applied in even, parallel passes, overlapping 50 percent.		
4.4.3.5a	Weather Conditions	[PCR 133] No coating shall be applied when contamination from any source (i.e., rainfall) is imminent or when the temperature or humidity is outside limits recommended by the coating manufacturer.		
4.4.3.5b	Weather Conditions	[PCR 134] To prevent condensation during application, the surface temperature shall be at least 3 °C (5 °F) above the dew point and rising.		
4.4.3.5c	Weather Conditions	[PCR 135] Spray application methods shall not be used when wind speed exceeds 25 km/hr (15 mph) in the area where the coating is being applied.		
4.4.3.5d(1)	Weather Conditions	[PCR 136] Solvent-based inorganic zinc coatings, polysiloxane topcoats, and IOTs shall not be applied in conditions with <40 percent RH.		
4.4.3.5d(2)	Weather Conditions	[PCR 137] Water-based inorganic zinc coatings shall not be applied in conditions with <40 percent or >80 percent RH.		
4.4.3.6a	Methods of Application	[PCR 138] Coatings shall be applied with airless or conventional spray equipment, or both, according to section 4.2.3.		
4.4.3.6b	Methods of Application	[PCR 139] Application with brushes shall be permitted for minor touchup of spray applications and stripe coats of inorganic zinc.		
4.4.3.7a	Coating Finish	[PCR 140] Each coat of material applied shall be free of runs, sags, blisters, bubbles, and mud-cracking; variations in color, gloss, and texture; holidays (missed areas); excessive film buildup; foreign contaminants; dry overspray; etc.		
4.4.3.7b	Coating Finish	[PCR 141] Special care shall be taken to ensure complete coverage and proper thickness on welds, corners, crevices, sharp edges, bolts, nuts, and rivets.		
4.4.3.7c	Coating Finish	[PCR 142] Each coat of applied material shall be rendered clean, dry, and free from surface contaminants before another coating is applied.		
4.4.3.8a	Touchup of Welds and Damaged Coatings	[PCR 143] Field welds and damaged coatings shall be touched up in accordance with section 4.5.8.		
4.4.3.8b	Touchup of Welds and Damaged Coatings	[PCR 144] The coating shall be applied in accordance with sections 4.4.3.4 and 4.4.3.6.		
4.4.3.8c	Touchup of Welds and Damaged Coatings	[PCR 145] Touchup and repair shall be accomplished promptly after the damage or welding has occurred.		

4.4.3.9a	Coating, Drying, and Curing	[PCR 146] The coating manufacturer's recommended drying and curing times for handling, recoating, and top coating shall be followed.		
4.4.3.9b	Coating, Drying, and Curing	[PCR 147] The coating manufacturer's recommendations shall be followed to test the coating for proper curing.		
4.4.3.9c	Coating, Drying, and Curing	[PCR 148] Proper curing of solvent-based inorganic zinc-rich coatings shall be verified by ASTM D4752, Standard Practice for Measuring MEK Resistance of Ethyl Silicate (Inorganic) Zinc-Rich Primers by Solvent Rub, prior to further coating.		
4.4.3.9d	Coating, Drying, and Curing	[PCR 149] Water-based inorganic zinc-rich coatings shall be verified for curing in accordance with the same procedure, but water is to be substituted as the solvent.		
4.4.4a	Sealing/Caulking	[PCR 150] The perimeter of all faying surfaces, joints open less than 13 mm (0.5 in), and skip-welded joints shall be completely sealed.		
4.4.4b	Sealing/Caulking	[PCR 151] The sealant shall be a self-curing, single-component, polysulfide rubber or polyurethane type, conforming to section 4.1.3.		
4.4.4c	Sealing/Caulking	[PCR 152] The sealant shall be applied to the joint with a caulking gun after the inorganic zinc primer has been applied on carbon steel.		
4.4.4d	Sealing/Caulking	[PCR 153] For topcoated zinc primers, caulking shall be applied after the intermediate coat of epoxy.		
4.4.4e	Sealing/Caulking	[PCR 154] For coatings on stainless steel, galvanized steel, and aluminum, caulking shall be applied before application of the topcoat.		
4.4.4f	Sealing/Caulking	[PCR 155] The bead shall have a smooth and uniform finish and be cured (tacky to the touch) before the topcoat is applied.		
4.5.1a	Protection of Carbon Steel	[PCR 156] Carbon steel surfaces shall be protected from atmospheric corrosion through the application of zinc coatings (inorganic zinc coating and/or hot-dip galvanizing and/or metallizing) as defined herein.		
4.5.1b	Protection of Carbon Steel	[PCR 157] New steel components, such as stair treads, grating, handrails, pipes, and hardware (nuts, bolts, and fasteners), shall be hot-dip galvanized in accordance with section 4.5.1.2.1, as applicable.		
4.5.1c	Protection of Carbon Steel	[PCR 158] All other carbon steel surfaces that are exposed to the atmosphere shall be coated with inorganic zinc conforming to section 4.1.2 in accordance with section 4.4.3, hot-dip-galvanized (zinc-coated) in accordance with section 4.5.1.2.1, or metallized in accordance with section 4.5.1.3.		
4.5.1d	Protection of Carbon Steel	[PCR 159] Carbon steel faying surfaces that are a part of all friction-type and electrical grounding joints shall be abrasive-blasted and coated with 100 µm to 150 µm (4 mil to 6 mil) of inorganic zinc only, in accordance with section 4.5.1.1.4, prior to installation.		
4.5.1e	Protection of Carbon Steel	[PCR 160] An inorganic zinc coating used in a friction-type joint shall be approved by the American Institute of Steel Construction (AISC).		

4.5.1f	Protection of Carbon Steel	[PCR 161] The recommended coating application sequence for carbon steel shall be to first abrasive blast the steel and then to prime it with inorganic zinc before installation or erection.		
4.5.1g	Protection of Carbon Steel	[PCR 162] Further topcoating, if required, shall be done after all welding, grinding, or drilling has been completed, and after areas damaged by these procedures have been properly repaired with inorganic zinc.		
4.5.1.1.1	Pre-Cleaning of Carbon Steel	[PCR 163] Carbon steel surfaces shall be cleaned and degreased in accordance with SSPC-SP 1 followed by power tool cleaning in accordance with SSPC-SP 3, Power Tool Cleaning, to remove weld spatter, weld slag, laminations, sharp edges, and other surface defects prior to abrasive blasting or power tool cleaning to bare metal.		
4.5.1.1.2a	Power Tool Cleaning of Carbon Steel	[PCR 164] Carbon steel shall be cleaned to bare metal, using power tools, in accordance with SSPC-SP-11, Power Tool Cleaning to Bare Metal, when a roughened, clean, bare metal surface is required but abrasive blasting is not feasible or permissible.		
4.5.1.1.2b	Power Tool Cleaning of Carbon Steel	[PCR 165] The surface anchor profile of the surface cleaned with the power tool shall be 40 μm to 75 μm (1.5 mil to 3.0 mil).		
4.5.1.1.2c	Power Tool Cleaning of Carbon Steel	[PCR 166] All rust shall be completely removed from pits and depressions.		
4.5.1.1.3a	Abrasive Blasting of Carbon Steel	[PCR 167] Carbon steel shall be abrasive-blasted to a minimum cleanliness of near-white metal, in accordance with SSPC-SP 10/NACE No. 2, Near-White Metal Blast Cleaning, with aggregate conforming to the requirements in section 4.1.1.		
4.5.1.1.3b	Abrasive Blasting of Carbon Steel	[PCR 168] The anchor profile of the blasted surface shall be 40 μm to 75 μm (1.5 mil to 3.0 mil), measured in accordance with ASTM D4417, Standard Test Methods for Field Measurement of Surface Profile of Blast Cleaned Steel.		
4.5.1.1.3c	Abrasive Blasting of Carbon Steel	[PCR 169] All rust shall be completely removed from pits and depressions.		
4.5.1.1.4	Stripe Coat Application	[PCR 170] Stripe coating with inorganic zinc shall be applied to welds, cutouts, sharp edges, rivets, crevices, and bolts to ensure complete coverage prior to subsequent applications of inorganic zinc.		
4.5.1.1.5a	Application of Inorganic Zinc Coatings	[PCR 171] Inorganic zinc coatings shall be applied to a DFT of 100 μm (4.0 mil) minimum to 150 μm (6.0 mil) maximum when they will be left without a topcoat or when IOT or ablative coating is applied.		
4.5.1.1.5b	Application of Inorganic Zinc Coatings	[PCR 172] When the zinc coatings are to be topcoated with organic topcoats, the DFT shall be reduced to 65 μm (2.5 mil) minimum to 100 μm (4.0 mil) maximum.		
4.5.1.1.5c	Application of Inorganic Zinc Coatings	[PCR 173] The proper DFT for the inorganic zinc coating shall be obtained in a single application, which may consist of multiple passes, while the coating is still wet (including the application of a stripe coat).		

4.5.1.1.6a	Topcoat Systems for Inorganic Zinc Coatings	[PCR 174] The following topcoat systems shall be applied over the inorganic zinc coatings as required for each zone of exposure described in section 1.4.		
4.5.1.1.6b	Topcoat Systems for Inorganic Zinc Coatings	[PCR 175] Topcoats shall be applied at the DFT recommended by the manufacturer or as specified in the section that follows.		
4.5.1.1.6c	Topcoat Systems for Inorganic Zinc Coatings	[PCR 176] The film thickness of the topcoats shall be sufficient to ensure uniform coverage and color.		
4.5.1.1.6c(2)	Topcoat Systems for Inorganic Zinc Coatings	[PCR 177] Zone 1c. Inorganic zinc coatings shall be left without a topcoat.		
4.5.1.1.6c(3)	Topcoat Systems for Inorganic Zinc Coatings	[PCR 178] Zone 2. An IOT conforming to section 4.1.2.3.3 shall be applied at a DFT of 75 μm to 125 μm (3 mil to 5 mil).		
4.5.1.1.6.c(4)A	Topcoat Systems for Inorganic Zinc Coatings	[PCR 179] Zones 3a and 3b. An intermediate/tie coat and a finish coat conforming to section 4.1.2 shall be applied in accordance with section 4.4.3.		
4.5.1.1.6.c(4)B	Topcoat Systems for Inorganic Zinc Coatings	[PCR 180] Zones 3a and 3b. As an alternate, an IOT conforming to section 4.1.2.3.3 or a polysiloxane finish coat conforming to section 4.1.2.3.4 shall be applied at the manufacturer's recommended DFT.		
4.5.1.1.6.c(4)C	Topcoat Systems for Inorganic Zinc Coatings	[PCR 181] Zones 3a and 3b. The DFT shall be sufficient to completely hide the inorganic zinc primer.		
4.5.1.1.6.c(5)	Topcoat Systems for Inorganic Zinc Coatings	[PCR 182] Zones 4a, 4b, and 4c. When required for color coding, safety purposes, identification, or special conditions, topcoats shall be in accordance with section 4.4.3.2.		
4.5.1.1.6c(7)	Topcoat Systems for Inorganic Zinc Coatings	[PCR 183] Zone 6. The coating system shall be as specified in sections 4.5.4 and 4.5.5.		
4.5.1.1.6c(8)	Topcoat Systems for Inorganic Zinc Coatings	[PCR 184] Zone 7. The coating system shall be as specified in NACE International RP0198, Control of Corrosion Under Thermal Insulation and Fireproofing Materials – A Systems Approach.		
4.5.1.2.1a	Galvanizing	[PCR 185] Galvanizing (zinc coating) shall be accomplished after fabrication by the hot-dip process conforming to ASTM A123, Standard Specification for Zinc (Hot-Dip Galvanized) Coatings on Iron and Steel Products; ASTM A153, Standard Specification for Zinc Coating (Hot-Dip) on Iron and Steel Hardware; and ASTM		

		A653, Standard Specification for Steel Sheet, Zinc-Coated (Galvanized) or Zinc-Iron Alloy-Coated (Galvannealed) by the Hot-Dip Process.		
4.5.1.2.1b	Galvanizing	[PCR 186] Galvanizing weight for steel sheet without further coating protection shall meet the standards of ASTM A653, with a galvanizing weight of G165.		
4.5.1.2.1c	Galvanizing	[PCR 187] All lower galvanizing weights for steel sheet shall be further protected with coatings except for Zone 5a and 5b exposures.		
4.5.1.2.1d	Galvanizing	[PCR 188] In accordance with this NASA Technical Standard, the galvanneal process shall not be used for the coating of steel sheet.		
4.5.1.2.1e	Galvanizing	[PCR 189] Steel components with an ultimate tensile strength above 900 MPa (130 ksi) or hardness above Rockwell C Hardness 28 shall not be galvanized due to potential hydrogen embrittlement.		
4.5.1.2.2a	Surface Preparation for Galvanizing	[PCR 190] Special care shall be taken to prevent any metal distortion by reducing blast nozzle pressure and increasing the working distance from the nozzle to the surface.		
4.5.1.2.2b	Surface Preparation for Galvanizing	[PCR 191] In some cases, such as in the surface preparation of light-gage sheet steel, these precautions may not be sufficient to prevent distortion; and alternate procedures, such as abrading or mechanical cleaning, shall be used to remove corrosion or roughen the surface.		
4.5.1.2.2c	Surface Preparation for Galvanizing	[PCR 192] Galvanized surfaces shall be abrasive-blasted with fine-grade abrasives conforming to the requirements in section 4.1.1 to remove corrosion and old coatings or roughen new surfaces.		
4.5.1.2.2d	Surface Preparation for Galvanizing	[PCR 193] The blasted surface shall be free of all corrosion and foreign matter and have a uniform, slightly roughened appearance.		
4.5.1.2.2e	Surface Preparation for Galvanizing	[PCR 194] Galvanized surfaces to be further topcoated shall be prepared by degreasing in accordance with section 4.4.2.1 before any additional surface preparation.		
4.5.1.2.2f	Surface Preparation for Galvanizing	[PCR 195] After degreasing, abrasive blasting or mechanical cleaning shall be performed as required by the zone of exposure, as defined in section 4.5.1.2.3.		
4.5.1.2.2g	Surface Preparation for Galvanizing	[PCR 196] If galvanized steel is prepared for the application of coatings by abrasive blasting, it shall be lightly brush-blasted with fine-grade abrasive at a lower pressure to provide a corrosion-free and uniform, slightly roughened surface.		
4.5.1.2.2h	Surface Preparation for Galvanizing	[PCR 197] Care shall be taken not to completely remove the galvanized finish.		
4.5.1.2.2i	Surface Preparation for Galvanizing	[PCR 198] The zinc coatings shall be maintained or rendered clean, dry, and free from contaminants before topcoat systems are applied.		
4.5.1.2.2j	Surface Preparation for Galvanizing	[PCR 199] Field repair of damaged galvanized surfaces shall be accomplished in accordance with ASTM A780, Standard Practice for Repair of Damaged and Uncoated Areas of Hot-Dip Galvanized Coatings, using inorganic zinc coatings.		

4.5.1.2.2k	Surface Preparation for Galvanizing	[PCR 200] Galvanized steel that is to be mechanically cleaned shall be cleaned in accordance with SSPC-SP 3 using abrasive discs/sheets, or other approved methods.		
4.5.1.2.2l	Surface Preparation for Galvanizing	[PCR 201] All corrosion and foreign matter shall be completely removed and the entire surface slightly roughened.		
4.5.1.2.3a	Coating Systems for Galvanizing	[PCR 202] Zones 1a and 1b. Galvanized surfaces may be left without a topcoat; however, for maximum protection, the galvanized coating shall be topcoated with a heat-resistant silicone ablative coating material in accordance with KSC-SPEC-F-0006.		
4.5.1.2.3b	Coating Systems for Galvanizing	[PCR 203] Zone 1c. Galvanized surfaces shall be left without a topcoat.		
4.5.1.2.3c	Coating Systems for Galvanizing	[PCR 204] Zone 2. After brush-blasting, an IOT conforming to section 4.1.2.3.3 shall be applied at a DFT of 75 μm to 125 μm (3 mil to 5 mil).		
4.5.1.2.3d	Coating Systems for Galvanizing	[PCR 205] Zones 3a and 3b. After brush-blasting, primer/tiecoat and finish coat conforming to section 4.1.2 shall be applied in accordance with manufacturer's recommended thicknesses.		
4.5.1.2.3d(1)	Coating Systems for Galvanizing	[PCR 206] Zones 3a and 3b. As an alternate, an IOT conforming to section 4.1.2.3.3 or a polysiloxane finish coat conforming to section 4.1.2.3.4 shall be applied at the manufacturer's recommended DFT.		
4.5.1.2.3d(2)	Coating Systems for Galvanizing	[PCR 207] Zones 3a and 3b. The DFT shall be sufficient to completely hide the galvanized coating.		
4.5.1.2.3.e.(1)	Coating Systems for Galvanizing	[PCR 208] Zones 4a, 4b, and 4c. When steel sheet is galvanized less than ASTM A653, with a galvanizing weight of G165, further coating in accordance with Zone 3 shall be required.		
4.5.1.2.3e(2)	Coating Systems for Galvanizing	[PCR 209] Zones 4a, 4b, and 4c. As an alternate to topcoats, steel sheet shall be degreased, brush-blasted, and an inorganic zinc primer conforming to section 4.1.2.1 applied to a DFT of 50 μm to 75 μm (2 mil to 3 mil).		
4.5.1.2.3f(1)	Coating Systems for Galvanizing	[PCR 210] Zones 5a and 5b. When topcoats are required for color coding, safety purposes, identification, or special conditions, the surface shall be degreased and an epoxy primer applied at the manufacturer's recommended DFT.		
4.5.1.2.3f(2)	Coating Systems for Galvanizing	[PCR 211] Zones 5a and 5b. Within 24 hours, a polyurethane finish coat conforming to section 4.1.2 shall be applied at the manufacturer's recommended DFT.		
4.5.1.2.3f(3)	Coating Systems for Galvanizing	[PCR 212] Zones 5a and 5b. As an alternate, a polysiloxane topcoat conforming to section 4.1.2.3.4 shall be applied at the manufacturer's recommended DFT.		
4.5.1.2.3f(4)	Coating Systems for Galvanizing	[PCR 213] Zones 5a and 5b. The DFT shall be sufficient to completely hide the galvanized coating.		

4.5.1.2.3g	Coating Systems for Galvanizing	[PCR 214] Zone 6. The coating system shall be as specified in sections 4.5.4 and 4.5.5.		
4.5.1.2.3h	Coating Systems for Galvanizing	[PCR 215] Zone 7. The coating system shall be as specified in NACE International RP0198.		
4.5.1.3.1	Pre-Preparation of Carbon Steel	[PCR 216] Carbon steel surfaces shall be cleaned and degreased in accordance with SSPC-SP 1 followed by power tool cleaning in accordance with SSPC-SP 3 to remove weld spatter, weld slag, laminations, sharp edges, and other surface defects prior to abrasive blasting or power-tool cleaning to bare metal.		
4.5.1.3.2a	Abrasive Blasting of Carbon Steel	[PCR 217] At a minimum, carbon steel shall be abrasive-blasted to near-white metal (SSPC-SP 10/NACE No. 2) with aggregate conforming to the requirements in section 4.1.1.		
4.5.1.3.2b	Abrasive Blasting of Carbon Steel	[PCR 218] The anchor profile of the blasted surface shall be 62.5 µm to 75 µm (2.5 mil to 3 mil).		
4.5.1.3.2c	Abrasive Blasting of Carbon Steel	[PCR 219] All rust shall be completely removed from pits and depressions.		
4.5.1.3.3a	Application of Metallized Coatings	[PCR 220] Metal wire to be used with the arc spray metallizing equipment shall be pure zinc, 90-10 zinc-aluminum, 85-15 zinc-aluminum alloys, 53-56 aluminum-magnesium, or pure magnesium.		
4.5.1.3.3b	Application of Metallized Coatings	[PCR 221] Metallized zinc coatings shall be applied to a DFT of 200 µm (8 mil) minimum to 375 µm (15 mil) maximum, depending on the intended service environment.		
4.5.1.3.4a	Topcoat Systems for Metallized Zinc Coatings	[PCR 222] Topcoat systems shall be applied over the metallized zinc coatings as required for each zone of exposure described in section 1.1.		
4.5.1.3.4.b	Topcoat Systems for Metallized Zinc Coatings	[PCR 223] The coating materials shall be selected from Appendix C, Approved Products List for Metallized (TSC) Systems.		
4.5.1.3.4c	Topcoat Systems for Metallized Zinc Coatings	[PCR 224] Topcoats shall be applied at the DFT recommended by the manufacturer or as specified in requirement "d" that follows.		
4.5.1.3.4d	Topcoat Systems for Metallized Zinc Coatings	[PCR 225] The film thickness of the topcoats shall be sufficient to ensure uniform coverage and color.		
4.5.1.3.4d(1)	Topcoat Systems for Metallized Zinc Coatings	[PCR 226] Zones 1a and 1b. Metallized coatings may be left without a topcoat; however, for maximum protection, the metallized coating shall be topcoated with a heat-resistant silicone ablative coating material in accordance with KSC-SPEC-F-0006.		

4.5.1.3.4d(2)	Topcoat Systems for Metallized Zinc Coatings	[PCR 227] Zone 1c. Metallized coatings shall be left without a topcoat.		
4.5.1.3.4d(3)	Topcoat Systems for Metallized Zinc Coatings	[PCR 228] Zone 2. An IOT conforming to section 4.1.2.3.3 shall be at a DFT of 75 µm to 125 µm (3 mil to 5 mil).		
4.5.1.3.4d(4)	Topcoat Systems for Metallized Zinc Coatings	[PCR 229] Zone 3. An intermediate/tie coat and a finish coat conforming to section 4.1.2 shall be applied in accordance with section 4.4.3.		
4.5.1.3.4d(4)A	Topcoat Systems for Metallized Zinc Coatings	[PCR 230] Zone 3. As an alternate, an IOT conforming to section 4.1.2.3 shall be applied at a DFT of 75 µm to 125 µm (3 mil to 5 mil), or a polysiloxane finish coat conforming to section 4.1.2.3.4 shall be applied at the manufacturer's recommended DFT.		
4.5.1.3.4d(4)B	Topcoat Systems for Metallized Zinc Coatings	[PCR 231] Zone 3. The DFT shall be sufficient to completely hide the metallized coating.		
4.5.1.3.4d(5)	Topcoat Systems for Metallized Zinc Coatings	[PCR 232] Zones 4a, 4b, and 4c. When topcoats are required for color coding, safety purposes, identification, or special conditions, topcoats shall be in accordance with section 4.4.3.2.		
4.5.1.3.4d(7)	Topcoat Systems for Metallized Zinc Coatings	[PCR 233] Zone 6. The coating system shall be as specified in sections 4.5.4 and 4.5.5.		
4.5.1.3.4d(8)	Topcoat Systems for Metallized Zinc Coatings	[PCR 234] Zone 7. The coating system shall be as specified in NACE International RP0198.		
4.5.2	Protection of Aluminum	[PCR 235] Aluminum shall be protected from corrosion by the use of protective coatings as defined herein.		
4.5.2.1a	Surface Preparation of Aluminum	[PCR 236] Special care shall be taken to ensure against any metal damage by the choice of abrasive aggregate and by reducing blast nozzle pressure and increasing the working distance from the nozzle to the surface as necessary.		
4.5.2.1b	Surface Preparation of Aluminum	[PCR 237] In some cases, such as in the surface preparation of light-gage sheet, these precautions may not be sufficient to prevent distortion; and an alternate procedure, such as abrading or mechanical cleaning, shall be used to remove corrosion or roughen the surface.		
4.5.2.1c	Surface Preparation of Aluminum	[PCR 238] Aluminum surfaces shall be abrasive-blasted with fine-grade abrasive materials conforming to the requirements in section 4.1.1 to remove corrosion and old coatings or roughen new surfaces.		

4.5.2.1d	Surface Preparation of Aluminum	[PCR 239] The blasted surface shall be free of all corrosion and foreign matter and have a uniform, slightly roughened appearance.		
4.5.2.1e	Surface Preparation of Aluminum	[PCR 240] Aluminum shall be prepared by degreasing and abrasive blasting or mechanical cleaning, as required by the condition and configuration of the surface.		
4.5.2.1f	Surface Preparation of Aluminum	[PCR 241] Abrasive blasting shall be used whenever possible using nonmetallic abrasives specified in section 4.1.1.		
4.5.2.1g	Surface Preparation of Aluminum	[PCR 242] Mechanical cleaning shall be used only when abrasive blasting is impractical, would damage the structure or component, or is prohibited in the area where the work is being performed.		
4.5.2.1h	Surface Preparation of Aluminum	[PCR 243] Aluminum shall be mechanically cleaned in accordance with SSPC-SP 3 using abrasive discs/sheets, or other approved methods.		
4.5.2.1i	Surface Preparation of Aluminum	[PCR 244] All corrosion and foreign matter shall be completely removed and the entire surface slightly roughened.		
4.5.2.1j	Surface Preparation of Aluminum	[PCR 245] Anodized or chemical-conversion-coated aluminum surfaces shall not be mechanically cleaned.		
4.5.2.1k	Surface Preparation of Aluminum	[PCR 246] In accordance with section 4.1.1, plastic media or an approved equivalent shall be used for abrasive blasting of bellows, gimbal joints, and other thin-walled, abrasion-sensitive components.		
4.5.2.2	Protective Coatings	[PCR 247] The following protective coatings shall be applied to aluminum surfaces as required for each zone of exposure described in section 1.4:		
4.5.2.2a	Protective Coatings	[PCR 248] Zones 1, 2, and 3. The following coatings shall be used to protect aluminum in the launch environment.		
4.5.2.2b	Protective Coatings	[PCR 249] Zones 4 and 5. Aluminum that is located within 3.5 km (2 mi) of the coastline or subject to chemical exposure shall be fully coated according to section 4.5.2.2.a.		
4.5.2.2c	Protective Coatings	[PCR 250] Zone 6. The coating system shall be as specified in sections 4.5.4 and 4.5.5.		
4.5.2.2d	Protective Coatings	[PCR 251] Zone 7. The coating system shall be as specified in NACE International RP0198.		
4.5.3a	Protection of Stainless Steel	[PCR 252] Type 300 series stainless steels shall be protected from corrosion by the use of protective coatings as defined in section 4.5.3.2.		
4.5.3b	Protection of Stainless Steel	[PCR 253] Note: *Thin-walled 300-series stainless-steel tubing is subject to pitting corrosion failure in outdoor marine environments.* For exterior installations, this tubing shall be degreased, prepared with a stainless-steel wire wheel or equivalent, and coated in accordance with section 4.5.3.2.		
4.5.3.1a	Surface Preparation of Stainless Steel	[PCR 254] Stainless steel shall be prepared by degreasing in accordance with SSPC-SP 1 and abrasive blasting or mechanical cleaning.		

4.5.3.1b	Surface Preparation of Stainless Steel	[PCR 255] Abrasive blasting shall be used whenever possible, using nonmetallic abrasives specified in 4.1.1.		
4.5.3.1c	Surface Preparation of Stainless Steel	[PCR 256] As an alternative, stainless steel shall be mechanically cleaned in accordance with SSPC-SP 3 using abrasive discs/sanding sheets or other approved methods.		
4.5.3.1d	Surface Preparation of Stainless Steel	[PCR 257] All corrosion and foreign matter shall be completely removed and the entire surface slightly roughened.		
4.5.3.1e	Surface Preparation of Stainless Steel	[PCR 258] Special care shall be taken to ensure against any metal damage by choice of abrasive aggregate and by reducing the blast nozzle pressure and increasing the working distance from the nozzle to the surface as necessary.		
4.5.3.1f	Surface Preparation of Stainless Steel	[PCR 259] In some cases, such as in the surface preparation of light-gage sheet, these precautions may not be sufficient to prevent distortion; and an alternate procedure, such as abrading or mechanical cleaning, shall be used to remove corrosion or roughen the surface.		
4.5.3.1g	Surface Preparation of Stainless Steel	[PCR 260] Stainless steel surfaces shall be abrasive-blasted with fine-grade abrasive conforming to the requirements in section 4.1.1 to remove corrosion and old coatings or roughen new surfaces.		
4.5.3.1h	Surface Preparation of Stainless Steel	[PCR 261] The blasted surface shall be free of all corrosion and foreign matter and have a uniform, slightly roughened appearance.		
4.5.3.2a	Protective Coating	[PCR 262] Zones 1, 2, and 3. For 300 series stainless steels, an inhibited polyamide epoxy primer and aliphatic polyurethane topcoat shall be used.		
4.5.3.2b	Protective Coating	[PCR 263] Zones 4 and 5. For special conditions, stainless steel shall be brush-blasted and coated with inhibitive epoxy primer to a DFT of 50 µm to 75 µm (2 mil to 3 mil) followed by a finish coat that provides a DFT of 50 µm to 75 µm (2 mil to 3 mil).		
4.5.3.2c	Protective Coating	[PCR 264] Zone 6. The coating system shall be as specified in sections 4.5.4 and 4.5.5.		
4.5.3.2d	Protective Coating	[PCR 265] Zone 7. The coating system shall be as specified in NACE International RP0198.		
4.5.4a	Underground, Submerged, or Continuously Wetted Surfaces	[PCR 266] Surfaces that will be underground, submerged, or continuously wetted shall be prepared in accordance with SSPC-SP 5/NACE No.1, White Metal Blast Cleaning, with a profile of 75 µm to 100 µm (3 mil to 4 mil) and coated with coal tar epoxy conforming to section 4.1.2.5.		
4.5.4b	Underground, Submerged, or Continuously Wetted Surfaces	[PCR 267] Coal tar epoxy coatings shall not be used for surfaces that will be in contact with potable water.		

4.5.4c	Underground, Submerged, or Continuously Wetted Surfaces	[PCR 268] The coating shall be applied to a minimum DFT of 410 μm (16.0 mil) and checked for missed areas or pinholes with a properly calibrated holiday detector in accordance with NACE International SP0188, Discontinuity (Holiday) Testing of New Protective Coatings on Conductive Substrates.		
4.5.4d	Underground, Submerged, or Continuously Wetted Surfaces	[PCR 269] Cathodic protection requirements shall be coordinated with the application of this coating.		
4.5.5a	Coating Systems for Potable Water Immersion Service	[PCR 270] All surface preparation for carbon steel shall be in accordance with SSPC-SP 5/NACE No. 1, with a surface profile of 75 μm to 100 μm (3 mil to 4 mil).		
4.5.5b	Coating Systems for Potable Water Immersion Service	[PCR 271] All coatings for potable water service shall be selected from section 4.1.2.6.		
4.5.5c	Coating Systems for Potable Water Immersion Service	[PCR 272] All potable water coating systems shall be inspected in accordance with standard recommended practices in NACE International RP0288, Inspection of Linings on Steel and Concrete, and with SP0188.		
4.5.6	Provision for Nonskid Surfaces	[PCR 273] Where a nonskid surface is required for walkways, decks, or other such surfaces, a nonskid coating conforming to section 4.1.2.7 shall be applied as follows:		
4.5.6a	Provision for Nonskid Surfaces	[PCR 274] Carbon steel. Coatings shall be applied directly over the zinc coating (inorganic zinc, galvanizing, or metallizing) and follow surface preparation instructions defined for topcoating in section 4.5.1.		
4.5.6b	Provision for Nonskid Surfaces	[PCR 275] Aluminum and stainless steel. Coatings shall be applied directly over these surfaces after surface preparation following instructions defined for topcoating in sections 4.5.2 and 4.5.3.		
4.5.7	Coating Systems for Metallic Surfaces Under Thermal Insulation	[PCR 276] Coating systems for carbon steel and stainless steel surfaces under thermal insulation and cementitious fireproofing shall be as specified in NACE International RP0198.		
4.5.8a	Repair of Applied Coatings	[PCR 277] Newly applied coatings shall be repaired in accordance with table 1, Repair of Applied Coatings. (Table 1 below)		

Table 1—Repair of Applied Coatings

Existing Coating	Repair Coating
Inorganic zinc	
Zones[1] 1 and 4	Inorganic zinc/epoxy mastic for small area touchup

Zone	Coating
Zone 2	Inorganic zinc/inorganic topcoat
Zones 3 and 5	Epoxy mastic/polyurethane/polysiloxane system for small area touchup
Galvanized steel	
Zones 1 and 4	Inorganic zinc/epoxy mastic for small area touchup
Zone 2	Inorganic zinc/inorganic topcoat
Zones 3 and 5	Epoxy mastic/polyurethane/polysiloxane system for small area touchup
Inorganic topcoat	
All zones	Inorganic zinc/inorganic topcoat
Epoxy/Polyurethane[2]	
Zones 3, 4, and 5	Epoxy/polyurethane system/polysiloxane
Water-reducible	
Zones 3, 4, and 5	Water-reducible intermediate/finish
Coal tar epoxy	
Zone 6	Coal tar epoxy

1 Zones are defined in section 1.4.
2 When this coating is replaced with inorganic zinc, complete removal of the existing coating is required.

4.5.8b	Repair of Applied Coatings	[PCR 278] Surfaces shall be prepared by washing with water and using mechanical methods in accordance with SSPC-SP 11, Power-Tool Cleaning to Bare Metal, to remove corrosion, weld slag, and to "feather back" coating edges.		
4.5.8c	Repair of Applied Coatings	[PCR 279] Touchup and repair shall be accomplished promptly after the damage has occurred.		
4.5.8d	Repair of Applied Coatings	[PCR 280] Touchup and repair of shop-applied coatings shall be accomplished using coatings from the same manufacturer as those applied in the shop.		
4.5.9a	Maintenance of Existing Coatings	[PCR 281] Each support contractor responsible for maintaining facilities or GSE shall develop a Coating Maintenance Plan that includes the following key elements: (1) Record keeping. (2) Routine inspection of facilities. (3) Coating repair criteria.		

		(4) Coating systems. (5) Equipment requirements. (6) Procedures. (7) Training and certification. (8) In-process inspection. (9) Worker protection. (10) Environmental compliance.		
4.5.9b	Maintenance of Existing Coatings	[PCR 282] All operations shall be in strict accordance with section 4.3.3.		
5.1.1	Responsibility for Inspection	[PCR 283] The coating contractor/applicator shall: a. Provide continuous quality control of all work to ensure complete conformance to the project specifications as defined in section 5.2. b. Submit a project-specific quality control coating inspection plan to the Contracting Officer for approval. c. Provide the NASA assigned coatings inspector with safe access to the work.		
5.1.2	Responsibility for Inspection	[PCR 284] The NASA-assigned coatings inspector shall be a NACE Certified Level III inspector under the NACE International Coating Inspector Program (CIP).		
5.1.3	Responsibility for Inspection	[PCR 285] Inspection of the surface preparation and coating application processes shall be performed by the NASA-assigned coatings inspector as follows: a. Perform all of the in-process inspections required by this NASA Technical Standard and the project specifications. b. Witness, inspect, and test all protective coating work to verify complete compliance with the specified requirements. c. Document the work on the inspection forms described in section 5.4. d. Prepare and sign the daily inspection reports on a daily basis and submit them to the Contracting Officer on a weekly basis as a minimum. e. When a nonconformance report is required, sign and submit it to the Contracting Officer within 1 workday from the time that it is written. f. After determining that all nonconformances have been corrected and/or the coating work is in compliance with this NASA Technical Standard and the project specifications, complete a conformance verification report for the specific item, area, or project. g. Sign and seal the conformance verification report. h. Not affix the seal to the daily inspection report or to the nonconformance report.		

5.2a	Requirements for Inspection	[PCR 286] Zones 1, 2, and 3. Since these zones are located in the highly corrosive launch environment or other chemical exposures, NACE inspection shall be required for all surface preparation and coating applications, including all new work, touchup of new work, major refurbishment of existing coatings, and modifications.		
5.2b	Requirements for Inspection	[PCR 287] Zone 4. For systems requiring abrasive blasting and coating of metallic substrates, all surfaces shall require full NACE inspection with the following exception: For touchup of existing coatings, NACE inspection is not mandatory but recommended in cases of critical systems or equipment.		
5.2c	Requirements for Inspection	[PCR 288] Zones 5a and 5b. All clean-room structures fabricated of aluminum or carbon steel that will be abrasive-blast-cleaned and/or coated outside Zone 5 environments shall require NACE inspection.		
5.2d	Requirements for Inspection	[PCR 289] Zone 6. Since this zone is located in a highly corrosive underground environment or other submerged exposures, NACE inspection shall be required for all surface preparation and coating applications, including all new work, touchup of new work, major refurbishment of existing coatings, and modifications.		
5.2e	Requirements for Inspection	[PCR 290] Zone 7. Since this zone is located in a highly corrosive environment, NACE inspection shall be required for all surface preparation and coating applications, including all new work, touchup of new work, major refurbishment of existing coatings, and modifications.		
5.3	Inspection Hold Points	[PCR 291] Mandatory inspection hold points shall include, but not be limited to, the following: a. Verification of ambient weather conditions in accordance with section 4.4.3.5. b. Prior to beginning of surface preparation work, to include the operation of equipment. c. After surface preparation work and before the beginning of the coating application work, to include the mixing of products. d. Before and after the application of each coat of material. e. After completion and prior to final acceptance.		
5.4	Inspection Forms	[PCR 292] All inspections shall be recorded and documented on forms acceptable to the customer.		
5.5	Inspection Prior to Surface Preparation and Coating Application	[PCR 293] The conditions in the following sections shall be inspected before beginning surface preparation and coating application operations.		
5.5.1a	Surface Condition	[PCR 294] The surface condition shall be visually inspected for compliance with section 4.4.2.		

5.5.1b	Surface Condition	[PCR 295] Special attention shall be given to weld spatter, sharp edges, flame or saw cuts, delaminations, burrs, slag, or other surface irregularities that affect performance of protective coatings prior to surface preparation.		
5.5.2a	Protection of Adjacent Surfaces	[PCR 296] Adjacent surfaces shall be visually inspected for adequate protection in accordance with section 4.4.2.		
5.5.2b	Protection of Adjacent Surfaces	[PCR 297] This inspection shall be jointly conducted with a Government Quality Engineering representative.		
5.5.3a	Ambient Weather Conditions	[PCR 298] The ambient weather conditions at the actual location of the work shall be determined before and during the surface preparation and coating application operations to ensure they are correct for the work being conducted.		
5.5.3b	Ambient Weather Conditions	[PCR 299] All measurement instrumentation shall be calibrated per the manufacturer's instructions prior to use.		
5.5.3c	Ambient Weather Conditions	[PCR 300] Proper instrumentation shall be used to measure air temperature, relative humidity, dewpoint, surface temperature, and wind speed and direction.		
5.5.3d	Ambient Weather Conditions	[PCR 301] No spray painting shall proceed when the measured wind speed in the immediate area of the coating work is above 25 km/hr (15 mph).		
5.5.3e	Ambient Weather Conditions	[PCR 302] All of these ambient weather conditions shall be recorded on the Coating System Daily Inspection Report as shown in Appendix G, Coating System Daily Inspection Report.		
5.5.4a	Compressed Air Cleanliness	[PCR 303] The compressed air supply shall be inspected for the use of inline moisture and oil traps.		
5.5.4b	Compressed Air Cleanliness	[PCR 304] Proper functioning of the traps shall be evaluated daily by allowing the air supply (down line from the traps) to blow against a clean, white cloth for several minutes, in accordance with ASTM D4285, Standard Test Method for Indicating Oil or Water in Compressed Air.		
5.5.5a	Surface Salt Concentration	[PCR 305] The surface chloride concentration shall be determined on all structures prior to surface preparation operations using an industry-recognized method, such as described in SSPC-TU 4, Field Methods for Retrieval and Analysis of Soluble Salts on Substrates, and recorded in the inspection records weekly.		
5.5.5b	Surface Salt Concentration	[PCR 306] Surfaces that measure 5 $\mu g/cm^2$ (0.00016 oz/ft^2) or above shall require washing with water in accordance with section 4.4.2.1 prior to surface preparation.		
5.6	Surface Preparation Inspection	[PCR 307] The inspections in the following sections shall be made to ensure compliance with the surface preparation requirements in section 4.4.2.		
5.6.1	Abrasive-Blasting Material	[PCR 308] The abrasive-blasting material shall be verified for compliance with section 4.1.1.		
5.6.2a	Blast Nozzle Air Pressure and Size	[PCR 309] The air pressure at the blast nozzle shall be determined through the use of a hypodermic needle air pressure gage.		

5.6.2b	Blast Nozzle Air Pressure and Size	[PCR 310] The needle of the gage shall be inserted as close to the nozzle as practically possible and in the direction of the air flow.		
5.6.2c	Blast Nozzle Air Pressure and Size	[PCR 311] Pressure readings shall be taken with the blasting system in full operation.		
5.6.2d	Blast Nozzle Air Pressure and Size	[PCR 312] The nozzle pressure shall be recorded.		
5.6.2e	Blast Nozzle Air Pressure and Size	[PCR 313] To ensure the compressor output correlates with the nozzle size, the nozzle shall be checked with a blast nozzle orifice gage initially and then at a frequency determined by the NACE inspector.		
5.6.3a	Degree of Surface Cleanliness	[PCR 314] The surface cleanliness shall be inspected after the surface preparation and before primer application to determine compliance with the applicable requirements of section 4.5.		
5.6.3b	Degree of Surface Cleanliness	[PCR 315] The degree of cleanliness of abrasive-blasted carbon steel shall be verified with a visual inspection in accordance with section 4.5.1.1.2.		
5.6.3c	Degree of Surface Cleanliness	[PCR 316] Galvanized steel, aluminum, and stainless steel shall be inspected for cleanliness in accordance with sections 4.5.1.2, 4.5.2, and 4.5.3.		
5.6.3d	Degree of Surface Cleanliness	[PCR 317] The surface preparation cleanliness requirements defined in section 4.5 shall be applicable to 100 percent of the subject area, including places that are difficult to reach.		
5.6.4a	Surface Profile or Roughness	[PCR 318] The anchor profile of an abrasive-blasted carbon steel surface shall be determined by using a surface profile gage, comparator, or replica tape.		
5.6.4b	Surface Profile or Roughness	[PCR 319] The profile shall be in accordance with section 4.5.1.1.2.		
5.6.4c	Surface Profile or Roughness	[PCR 320] Galvanized steel, stainless steel, and aluminum surfaces shall be visually inspected as required for slight roughening in accordance with sections 4.5.1.2, 4.5.2, and 4.5.3.		
5.6.5a	Blasting of Abrasive-Sensitive Components	[PCR 321] Thin-walled, abrasive-sensitive components, such as bellows assemblies or tubing, shall be protected during normal blasting operations in accordance with section 5.5.2.		
5.6.5b	Blasting of Abrasive-Sensitive Components	[PCR 322] Walnut shells or an approved equivalent shall be used for surface preparation of these sensitive components in accordance with section 4.1.1 or mechanical methods in accordance with section 4.4.2.3.		
5.7	Coating Application Inspection	[PCR 323] The inspections in the following sections shall be made to ensure compliance with the coating application requirements defined in section 4.4.3.		
5.7.1a	Surface Condition	[PCR 324] The prepared surface shall be visually inspected.		
5.7.1b	Surface Condition	[PCR 325] The time before coating shall be monitored for compliance with section 4.4.3 before coatings are applied.		

5.7.2	Coating Materials	[PCR 326] The coating materials shall be visually inspected for compliance with section 4.4.3.1.		
5.7.3	Storage of Coating Material	[PCR 327] Coating material storage conditions shall be periodically inspected for compliance with section 4.4.3.3.		
5.7.4	Mixing and Application of Coatings	[PCR 328] The mixing and application of all coatings shall be visually inspected to ensure compliance with sections 4.4.3.4, 4.4.3.6, and 4.4.3.9.		
5.7.5a	Coating Finish and DFT	[PCR 329] The finish and DFT of each applied coating shall be inspected for compliance with sections 4.4.3.7 and 4.5 prior to the application of successive coats.		
5.7.5b	Coating Finish and DFT	[PCR 330] The DFT measurement on carbon steel shall be taken using a magnetic gage calibrated in accordance with SSPC-PA 2, Procedure for Determining Conformance to Dry Coating Thickness Requirements.		
5.7.5c	Coating Finish and DFT	[PCR 331] DFT measurements on aluminum and stainless steel shall be taken using an eddy current instrument that has been properly calibrated on surfaces similar to the coated surface.		
5.8	Caulking Inspection	[PCR 332] All surfaces shall be visually inspected to determine whether they comply with the requirements for sealing and caulking in accordance with section 4.4.4.		
5.9	Galvanizing Inspection	[PCR 333] Galvanized carbon steel shall be inspected in accordance with the applicable ASTM standard in section 4.5.1.2.1.		

www.ingramcontent.com/pod-product-compliance
Lightning Source LLC
Chambersburg PA
CBHW081736220526
45468CB00008B/2116

* 9 7 8 1 7 9 5 6 5 1 1 1 0 *